Innovation Heroes

Understanding Customers as a Valuable Innovation Resource

Series on Technology Management

Series Editor: J. Tidd (*University of Sussex, UK*) ISSN 0219-9823

The Technology Management Series is dedicated to the advancement of academic research and management practice in the field of technology and innovation management. The series features titles which adopt an interdisciplinary, multifunctional approach to the management of technology and innovation, and includes work which seeks to integrate the management of technological, market and organisational innovation. All titles are based on original empirical research, and includes research monographs and multiauthor edited works. The focus throughout is on the management of technology and innovation at the level of the organisation or firm, rather than on the analysis of sectoral trends or national policy.

Published

Vol. 31 *Innovation Heroes: Understanding Customers as a Valuable Innovation Resource*
by Fiona Schweitzer (Grenoble École de Management, France & University of Applied Sciences Upper Austria, Austria) & Joe Tidd (University of Sussex, UK)

Vol. 30 *Innovation Discovery: Network Analysis of Research and Invention Activity for Technology Management*
edited by Tugrul Daim (Portland State University, USA) & Alan Pilkington (University of Westminster, UK)

Vol. 29 *Exploiting Intellectual Property to Promote Innovation and Create Value*
edited by Joseph Tidd (University of Sussex, UK)

Vol. 28 *Promoting Innovation in New Ventures and Small- and Medium-Sized Enterprises*
edited by Joseph Tidd (University of Sussex, UK)

Vol. 27 *The Role of Creativity in the Management of Innovation:State of the Art and Future Research Outlook*
edited by Alexander Brem (University of Southern Denmark, Denmark), Rogelio Puente-Diaz (University Anáhuac, Mexico) & Marine Agogué (HEC Montréal, Canada)

Vol. 26 *Total Value Development: How to Drive Service Innovation*
by Frank M. Hull (Cass Business School and Fordham University Graduate School of Business-Executive Education, USA) & Christopher David Storey (University of Sussex, UK)

Vol. 25 *Small Firms as Innovators: From Innovation to Sustainable Growth*
by Helena Forsman (University of Tampere, Finland)

More information on this series can also be found at http://www.worldscientific.com/series/stm

(Continued at the end of the book)

SERIES ON TECHNOLOGY MANAGEMENT – VOL. 31

Innovation Heroes

Understanding Customers as a Valuable Innovation Resource

Fiona Schweitzer
Grenoble École de Management, France
University of Applied Sciences Upper Austria, Austria

Joe Tidd
Science Policy Research Unit (SPRU), University of Sussex, UK

World Scientific

NEW JERSEY • LONDON • SINGAPORE • BEIJING • SHANGHAI • HONG KONG • TAIPEI • CHENNAI • TOKYO

Published by

World Scientific Publishing Europe Ltd.

57 Shelton Street, Covent Garden, London WC2H 9HE

Head office: 5 Toh Tuck Link, Singapore 596224

USA office: 27 Warren Street, Suite 401-402, Hackensack, NJ 07601

Library of Congress Cataloging-in-Publication Data
Names: Schweitzer, Fiona, author. | Tidd, Joseph, 1960– author.
Title: Innovation heroes : understanding customers as a valuable innovation resource /
Fiona Schweitzer (Grenoble École de Management, France & University of Applied Sciences Upper Austria, Austria), Joe Tidd (University of Sussex, UK).
Description: New Jersey : World Scientific, [2018] |
Series: Series on technology management ; Volume 31
Identifiers: LCCN 2018002462 | ISBN 9781786345363 (hc : alk. paper)
Subjects: LCSH: Customer relations. | Consumers. | New products. | Diffusion of innovations. |
Technological innovations.
Classification: LCC HF5415.5 .T493 2018 | DDC 658.8/12--dc23
LC record available at https://lccn.loc.gov/2018002462

British Library Cataloguing-in-Publication Data
A catalogue record for this book is available from the British Library.

Infographics design by Herbert Gsottbauer
Email: hgsottbauer@gmail.com
https://www.linkedin.com/in/herbert-gsottbauer/en

First published 2018 (Hardcover)
Reprinted 2019 (in paperback edition)
ISBN 978-1-78634-781-7 (pbk)

For any available supplementary material, please visit
https://www.worldscientific.com/worldscibooks/10.1142/Q0158#t=suppl

Desk Editors: Sundar Madhan/Jennifer Brough/Koe Shi Ying

Typeset by Stallion Press
Email: enquiries@stallionpress.com

Preface

"We have already tried it and it won't work." "Integrating customers does not work for our radical innovation because customers are unable to conceptualize or evaluate ideas beyond the realm of their immediate own experience." We have heard these and similar quotes of disillusioned R&D practitioners who asked us for advice on how to make customer integration work for them. But we also heard: "If our customer would not have mentioned this, we would have lost out an unfathomable new growth opportunity."

Notably in high-tech industrial goods markets, R&D professionals often have an ambiguous attitude toward customer integration. On the one hand, they are fascinated by the opportunities that customer integration can bring to their innovation projects: better customer insights inspire the creative process, ideas from knowledgeable customers propel explorative R&D activities, and the optimization of innovations facilitates meeting unsatisfied customer needs. At the same time, managers feel insecure when it comes to designing customer integration projects and applying customer integration techniques successfully.

We worked together with many of these companies to unravel the myths that exist internally in respect of customer integration and to provide them with the means to carry out customer integration activities effectively and efficiently. This book provides condensed insights into the key learnings we gathered over the years of practical experience as well as from our research emphasis in the area. We hope that the book serves as a fruitful basis for practitioners.

We designed the book as a useful and practical guide through the jungle of customer integration methods. It intends to provide a quick overview on the key principles and tools and clear explanation of customer integration methods. Case applications and practical hints make it a hands-on support tool that should reduce the barrier toward trying out the one or the other method for one's own innovation project.

The first feedback we received from practitioners who read the book indicates that we reached this goal:

One of the must reads when it comes to new product development! This book is well structured and provides an excellent overview on the various methods available to improve new product development by integrating customers into various stages of the new product development process. It discusses traditional customer integration methods as well as modern ICT-enabled ones in a clear and comprehensible way. Much more than this, it gives needed answers for a successful practical implementation.

Mag. Martin Kronthaler, Strategic Product Management Viking,
Stihl International GmbH

Customer integration and the consideration of customers' requirements are nothing new, but nowadays approaches are different and also more complex. The integration is much deeper and more systematic and covers the whole development process. Especially for a machine manufacturer new approaches and methods are described within this book in a very expedient way enabling us to use the potential of customers' know-how for our product and service developments.

Dr. Gerhard Dimmler, Senior Vice President of R&D for Products
at Engel Austria GmbH in Schwertberg, Austria

We wish an enjoyable and meaningful read.

Fiona Schweitzer
Joe Tidd

About the Authors

Fiona Schweitzer is a professor of marketing at Grenoble École de Management and at the University of Applied Sciences Upper Austria. Her academic research focuses on visualization in concept tests, technology adoption, social innovation, and customer integration into the innovation process. She has won three best-paper awards and has presented her work at various scientific conferences and published them as book chapters and in academic journals such as *Journal of Product Innovation Management*, *International Journal of Innovation Management*, *Psychology & Marketing*, *Creativity & Innovation Management*, and *Research-Technology Management*. She regularly works with high-tech companies such as Wacker Neuson, Siemens, and Viking/Stihl on diverse marketing and innovation management tasks. She worked as a visiting professor at the London Business School, the University of Berkeley, and the University of Hong Kong and gathered practical innovation and marketing skills at B2C companies (such as Unilever) and NGOs (such as WWF).

Joe Tidd is a physicist with subsequent degrees in technology policy and business administration. He is a professor of technology and innovation management at SPRU, visiting professor at University College London, and former professor of Cass Business School, Copenhagen Business School, and Rotterdam School of Management. He is the former deputy director of SPRU and head of the Innovation Group and director of the Executive MBA Program at Imperial College. He has worked as a policy adviser to the Confederation of British Industry (CBI). He was a researcher for the 5-y International Motor Vehicle Program of the Massachusetts Institute of Technology (MIT) and has worked on technology and innovation management projects for consultants Arthur D. Little, CAP Gemini, and McKinsey and numerous technology-based firms. He has written nine books and more than 60 papers on the management of technology and innovation, with more than 17,000 research citations.

Contents

Chapter 1

Customers at the center stage

1.1 Introduction

Companies are increasingly opening up their innovation processes to customers. Customer integration could boost new product and service success, as it increases the product's/service's fit to customer needs and wants. For example, Threadless, a US-based T-shirt manufacturer allows its customers to preselect the T-shirt designs it produces. Customer integration also provides easy accesses to customers' creative talents as an inspiration for new product and service ideas or as a contribution to problem solving throughout new product and service development (NPD and NSD). Local Motors unchains users' technical and design-related expertise by allowing them to codesign innovative car models. Lego, the manufacturer of small play bricks, runs the vibrant virtual community platform Lego Ideas to which users can upload their sketches of brick sets. The Lego management board considers producing sketches that receive a minimum of 10,000 votes. When this happens, the user whose idea it was receives 1% of the product net sales. The successful Lego Mindcraft line originates from this community.

Customer integration is not only a business-to-consumer (B2C) topic, but business-to-business (B2B) companies also strive to exploit its results. 3M has successfully carried out lead user workshops to codevelop concepts in the fields of medical imaging, surgical drapes, and infection control.

This company has also rolled out the lead user method beyond the medical division (von Hippel *et al.*, 2000; von Hippel, 2005). IBM runs Industry Solution Labs in four locations (New York, Zurich, New Delhi, and Beijing). The labs are the starting point for innovative partnership projects, connecting future client needs with emerging technologies. In these labs, IBM discusses individual client companies' complex business challenges, and key technology trends that IBM has captured in its annual Global Technology Outlook report, which serves as a basis for strategic planning inside IBM. The discussion of the Global Technology Outlook focuses on the impact of trends on clients' business, as well as on changes in trends and new trends, which will subsequently serve as a basis for the following year's report. The Industry Solution Labs also function as a starting basis for first-of-a-kind projects where IBM scientists and clients test a new technology on a real business problem for the first time. This process allows them to leapfrog the traditional development cycle and helps guide research efforts toward strategic markets. IBM and the client share the costs and risks, while client benefits from the competitive advantage gained by being the first to use the new technology (Kaiserswerth, 2013).

Where there is light, however, there is also shadow. Several innovation practitioners and researchers complain that customers are conservative and find it difficult to visualize future product usage when confronted with radical innovations. They thus fail to evaluate new product concepts correctly, let alone generate meaningful new product ideas.

Seizing the opportunities of customer integration while avoiding the pitfalls and negative side effects requires a clear understanding of customer integration's goals and of the value of including different customer types prior to choosing the customer integration method. A first approach to thinking about customer integration goals is finding answers to questions such as the following: What output do you expect from the customer integration activity? Are you looking for radically new products and services or incremental ones? Are you searching for a solution to a tricky technical problem? Are you searching for inspirations on a potential customer problem that you can solve through a new product, service, or product–service combination? Are you trying to identify latent customer needs? Do you want to assess your concepts' and ideas' perceived customer value in order to determine their market acceptance?

1.2 Opening and closing the solution space

Questions about the customer integration's goal are closely connected to the problem-solving tasks that NPD teams perform. These teams acquire customer knowledge in an endeavor to solve creative problems (Rindfleisch and Moorman, 2001; Parry and Song, 2010). In the NPD process, creative problem solving can be differentiated into divergent thinking and convergent thinking tasks. In divergent thinking, a person generates a number of ideas. This thinking style is associated with intuition, spontaneity, taking paradoxical perspectives, breaking with established thinking patterns, as well as unconventional and nonlinear problem solving. This style provokes out-of-the-box thinking, encourages the breaking of existing frames of reference, and aims at generating a large quantity of original solutions. In convergent thinking, a person applies logical, linear, and systematic patterns and steps to eliminate ambiguities and arrive at a single best answer to a problem. It is associated with critical thinking and using probabilities and standards to arrive at conclusions (Guilford, 1967).

In NPD, divergent thinking tasks focus on opening the solution space by identifying significant new customer problems and opportunities, generating new product and service ideas and concepts, and searching for technical possibilities to turn ideas into producible goods. Opening the solution space aims at generating multiple perspectives, as well as novel and unique ideas and solutions. Once the open solution space has generated enough new input, closing it requires evaluations and detailed considerations in order to select the best option. Convergent thinking closes the opportunity space by prioritizing opportunities, ideas, and concepts on the basis, for instance, of their meaningfulness and novelty (Isaksen *et al.*, 1994; Proctor, 2013).

The two tasks of opening and closing the solution space are an excellent basis for categorizing the vast array of customer integration methods. Figure 1 illustrates the different roles that customer involvement plays in opening and closing the NPD solution space.

In terms of opening the solution space, customer integration aims to gather customer input in order to develop various options regarding strategic arenas, product and service ideas, and concepts. Managers can integrate their customers directly or indirectly into activities aimed at opening

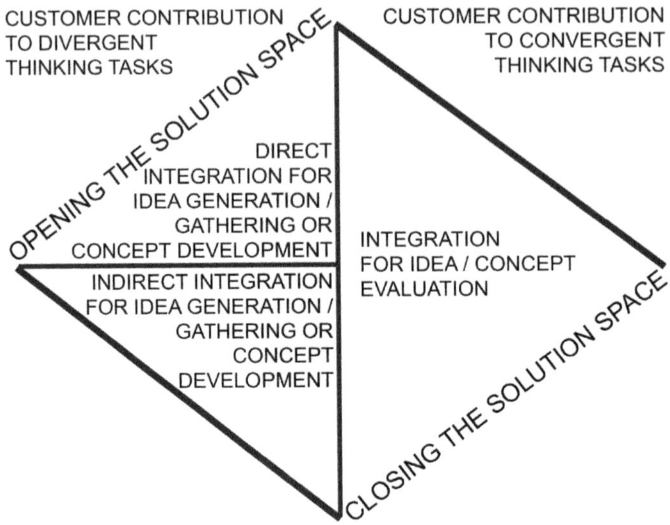

Figure 1: Customer integration to acquire information for opening and closing the NPD solution space.

the solution space. Direct activities include methods that aim at directly generating ideas, developing concepts, or codeveloping products and services with customers. In direct methods, customers' role is that of solution providers. In this category, typical methods used are online idea competitions, virtual idea communities, R&D alliances with customers, and lead user workshops.

In indirect methods, the interaction with customers inspires NPD/NSD teams' development efforts. Customers function not as codevelopers in indirect methods but as providers of latent needs information. These methods thus help NPD/NSD teams decipher unexpressed needs and translate them into new products and services. Ethnographic approaches, outcome-driven innovation (ODI), and the repertory grid technique are all indirect customer integration methods. A case (Rosenthal and Capper, 2006) from Panasonic's household electronic devices division illustrates the use of ethnography to detect latent needs. This case involved the electric shaver market for females and led to the NPD/NSD teams using novel materials to redesign the company's shavers. For example, the observation of users in spatially restricted, wet-shaving environments led to ergonomic design changes, including a shift from shiny, hard-edged form to soft, curvilinear ones.

In terms of closing the solution space, customer integration helps managers select, from a set of strategic arenas, product and service ideas, or concepts the one or the few worth carrying forward through the NPD/NSD process. In closing tasks, customers evaluate and choose from options. Customer integration methods for closing the solution space fall into two subcategories: qualitative and quantitative methods. Concept testing is a classic example of such methods. Qualitative methods are rich in verbal feedback rather than numbers. They allow for a rich understanding of customer evaluations' underlying reasons and motivations throughout the NPD/NSD process. Qualitative concept tests reveal why the target customers like or dislike a product or service, provide in-depth knowledge of the changes they would like to see in product or service features, as well as the reasons for requesting these changes. Data is gathered by means of semistructured personal interviews or group discussions employing open-ended questions, while the data analysis is nonstatistical. Quantitative methods provide descriptive data from large numbers of representative cases, thus allowing conclusions to be drawn from the sample to the target population. Quantitative concept testing employs both strictly structured surveys to gather data and statistical data analyses in order to gather adoption information from respondents, which help forecast their purchase intention and the market demand of the key markets' different market segments.

Figure 2 categorizes the various customer integration methods according to their ability to provide insights into their ability to directly or indirectly open and close the solution space via qualitative or quantitative approaches. Some methods, such as focus groups, are very versatile and flexible; they are therefore found in more than one of the four categories. Focus groups can be adapted to fit opening and closing tasks. Creativity sessions in focus group workshops entice customers to generate ideas, while questions on personal experiences with products and services stimulate revelations of their personal background, attitudes, and behaviors, thus resulting in knowledge about customers' latent needs. In addition, a shift to concrete feedback on the presented concepts provides information on customers' purchase intention and the factors influencing this intention.

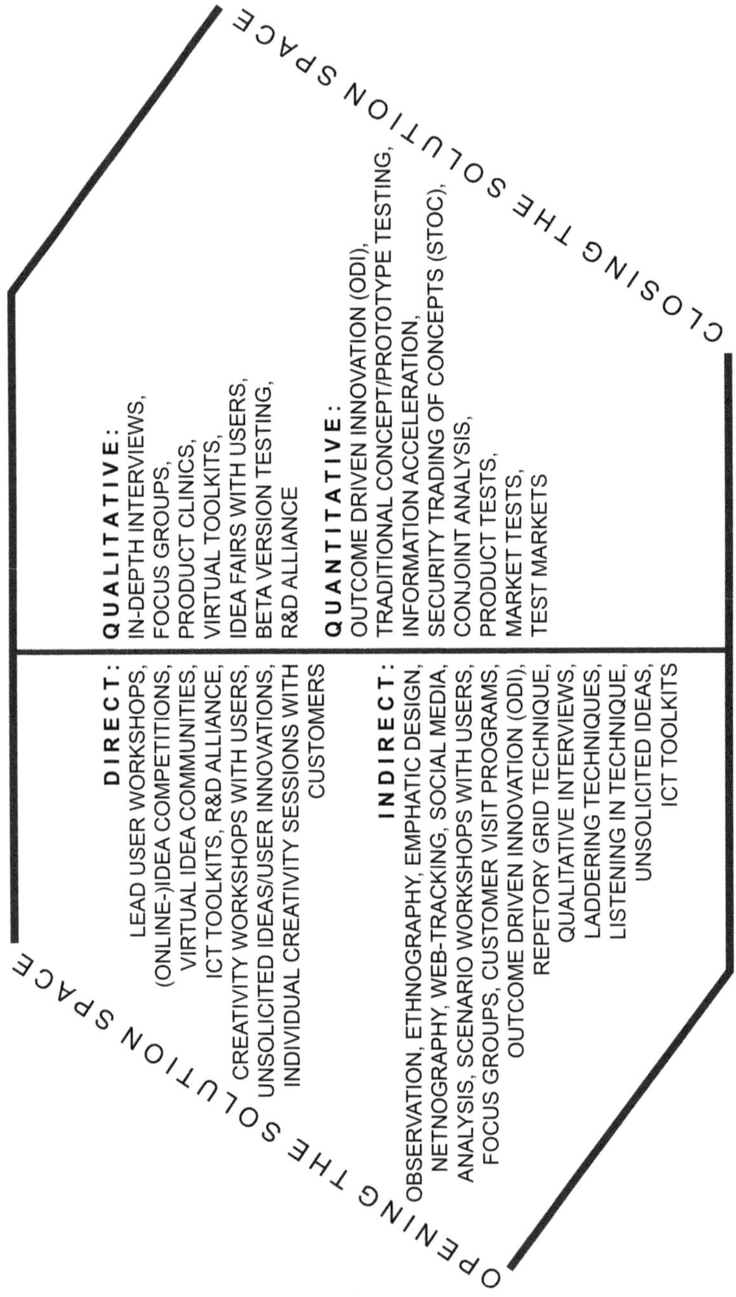

Figure 2: Customer integration tools for opening and closing the solution space.

Other methods fit one task specifically. For example, the repertory grid technique is limited to those methods that identify latent needs. Other methods are stepwise processes that form a combination of opening and closing tasks. For example, ODI comprises specific qualitative interviewing techniques involving a small number of customers to elicit their latent needs first. In a subsequent quantitative survey, customers are asked to evaluate the identified latent needs, which forms the basis for NPD/NSD teams to search for solutions that satisfy the most relevant latent needs in successive steps.

The categorization of methods according to their ability to provide direct or indirect customer insights into opening the solution space, as well as qualitative or quantitative information on closing the solution space, helps choose appropriate methods for different customer integration tasks. This categorization allows a three-step procedure for successful customer integration in NPD/NSD: First, the innovating company clearly defines the goals of the customer integration activity. Second, it chooses the right category of customer integration method based on these goals. Third, it selects a specific method within the category based on the advantages and disadvantages of the optional methods and the innovating company's experience with customer integration activities.

1.3 Customer integration along the different NPD phases

The NPD process covers a bevy of activities that an NPD team undertakes to transform rough ideas into real products for a market launch. The core idea of NPD processes is to launch successful products by continuously filtering bad ideas and refining good ones, eventually resulting in the realization of the most promising ones. The first approaches to structure NPD into consecutive subtasks and to review the points that allow managers to discern between successful and ineffective ideas can be traced back to the 1970s (Utterback, 1971; Utterback and Abernathy, 1975).

The stage gate model (Cooper, 1990, 1994) is an important NPD process model that is widely used in management practice (Cooper and Kleinschmidt, 1991; Griffin, 1997). The process guides NPD activities through a series of prescribed stages with each stage's achievements evaluated at the gates. In these gates, projects are prioritized; those that do not meet the gate criteria are filtered out and the remaining are moved to the subsequent, costlier stage. The stages evolve from early idea generation to concept design, product development, testing, validation, and launch (Cooper, 2008). Throughout each stage, NPD teams not only undertake tasks to open the solution space but also carry out closing tasks to prepare for the gate evaluations.

Prior research has demonstrated that stage gate processes improve top management's ability to exert control over NPD (Poskela and Martinsuo, 2009; Rijsdijk and van den Ende, 2011; Schultz *et al.*, 2013), entail more discipline regarding go/kill decision-making (Cooper, 2008; Schultz *et al.*, 2013), and speed up a firm's NPD efforts (Shaw *et al.*, 2001; Salomo *et al.*, 2007). While stage gate processes introduce a certain level of rigor and control to NPD processes, which helps reduce the uncertainties and increase the speed (Griffin, 1997; Cooper *et al.*, 2002), they also have drawbacks. Increased control comes at the expense of flexibility (Sethi and Iqbal, 2008; Lenfle and Loch, 2010). Since certain parameters are frozen after a project passes a specific gate, gates undermine the NPD team's ability to effect changes as they see fit (Verganti, 1999). Sethi and Iqbal (2008) provide empirical evidence of the negative effect of such inflexibility on NPD teams' ability to gather and integrate new information. To cure these shortcomings, advanced stage gate models include spiral and agile processes that allow repetitive, iterative feedback loops (Cobb, 2011; Conforto and Amaral, 2015; Cooper and Sommer, 2016), as well as overlapping stages and contingent gates that allow concurrent activities (Cooper, 1994; Ettlie and Elsenbach, 2007).

Irrespective of gate rigor and iteration options, any NPD process includes certain activities for opening and closing the solution space around the raw idea, the product concept, and the final product. Customers can be integrated into all these opening and closing activities to incorporate the customer perspective. This step promises to reduce the technical and market uncertainties throughout the NPD process and to provide the insights needed to make the right managerial decisions (Figure 3).

Strategic arenas connect strategic innovation foresight work with its operational execution in the NPD process. Strategic arenas set the

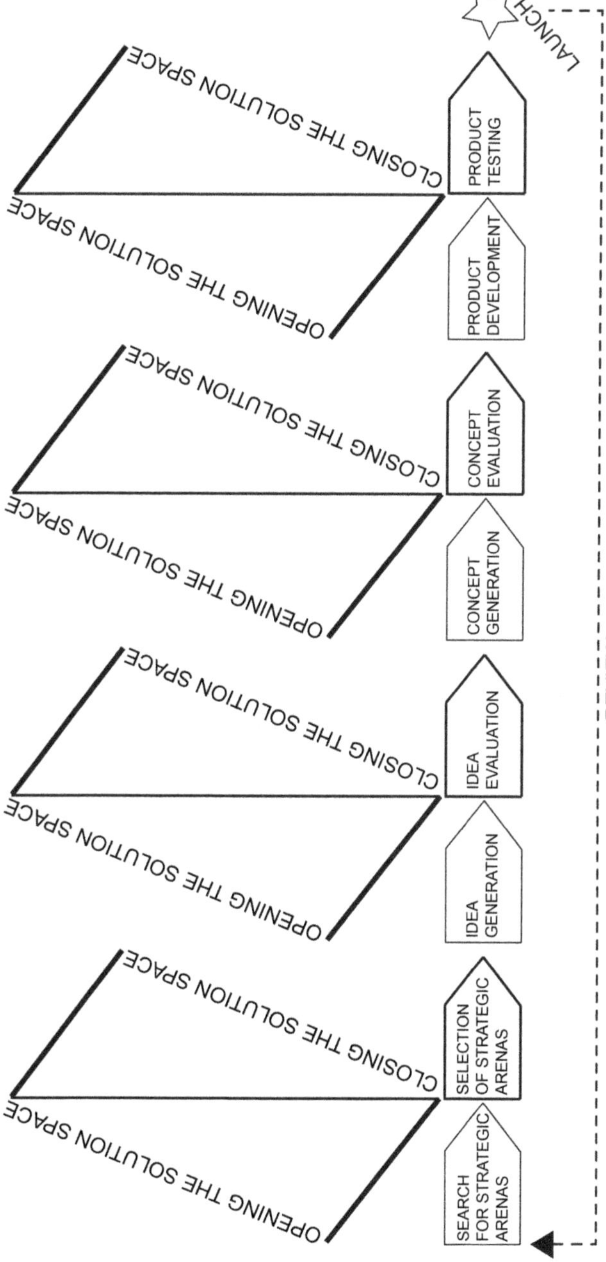

Figure 3: Customer integration throughout NPD.

framework within which the operational execution takes place, as they direct idea generation and search to a specific focus field. Strategic arenas thus connect idea-initiating insights into markets or technologies with idea-generation processes (Hambrick and Fredrickson, 2005).

For example, market insight may stem from a customer survey that reveals the customers' wish for improved safety in cars. Accident prevention (active safety) and a reduction of the consequences of accidents (passive safety) are strategic arenas that could help increase safety. A company can generate different solutions to increase car safety in each of these arenas. In other words, the search for strategic arenas does not yet focus on product solutions but constitutes a search for fields in which corporate competences can be utilized to generate future growth (search for strategic arenas), followed by the closing of the solution space through the selection of the strategic arenas with the highest leverage for goal achievement. An inappropriate strategic arena setting runs the risk of creating a gap between the strategic foresight and the start of NPD processes. Customer integration allows for gauging new technological insights' market relevance, generating market insights for a strategic arena setting, and for selecting strategic arenas from a market perspective. Methods of indirect customer integration through latent needs detection provide important information for this phase. For example, voestalpine Stahl, a leading Austrian manufacturer of high-quality steel, carried out a 3-day future workshop with a key client company's interdisciplinary team and its own employees. This conference was a central element in the definition of strategic arenas. Simultaneously, it led to a joint vision of certain strategic arenas' relevance and fostered the ambition to jointly work toward developing new products for these strategic arenas.

As soon as fertile strategic arenas are selected, two approaches can be used to search for new product ideas within them. One is idea gathering, which describes the search for existing ideas from different internal and external sources (e.g., employees, customers, competition), but not activities for producing ideas. Idea generation, the second approach, actively employs creativity methods to generate new approaches to solving a problem in individual or group creativity sessions. Beside these two extreme approaches, hybrid forms use elements of both. In this vein, online idea competitions gather external ideas and usually do not involve creativity techniques. Nevertheless, these competitions require the participants to not only propose existing ideas but also generate new ones, while the manufacturer sets the theme for idea generation and provides tools to facilitate idea transfer.

Customer integration in action: User-generated hits at Muji

Muji is a Japanese manufacturer and retailer with stores in 22 countries. It produces a broad range of consumer goods, focusing on interior and household products. 20% of its sales stems from furniture. The company has a reputation for innovative design and has won more than 90 design and innovation prizes. The furniture division started experimenting with user-generated products in 2002, when it launched "Floor Sofa," a soft upholstery that adjusts smoothly to the body. As this was a new approach, the company chose to reduce its risks through binding customer preorders before investing in the sofa's manufacturing. Floor Sofa sales reached staggering 500 million yen (equivalent to EUR 4 million) in the first year — one of the most successful launches in Muji's history. In 2003, a user-generated movable wall was equally successful. Spurred by its success, Muji planned a more systematic customer integration into NPD. From 2005 to 2009, Muji launched 43 new products, six of which were user generated via their idea website muji.net. Muji specified themes, such as "Suwaru Seikatsu" ("Sit-Down Life"), to generate ideas on specific topics. In the first phase, users are asked to submit ideas; in the second step, they have the opportunity to comment on ideas. Muji selects

(Continued)

(Continued)

the most promising ideas, refines them internally (e.g., by choosing fillings or fabrics), and adapts them for production. Muji employees are heavily involved in the last stage of the NPD of realizing the product. This effort is comparable to that in nonuser-generated products.

The unit sales and sales revenues' gross-margins in the first year, as well as the aggregate 3-y data, outperformed those of nonuser-generated products (unit sales results provided in Figure 4). Key decision-makers inside Muji also rated these products' novelty and strategic importance higher than those of nonuser-generated products, as captured by their potential to generate a new product line.

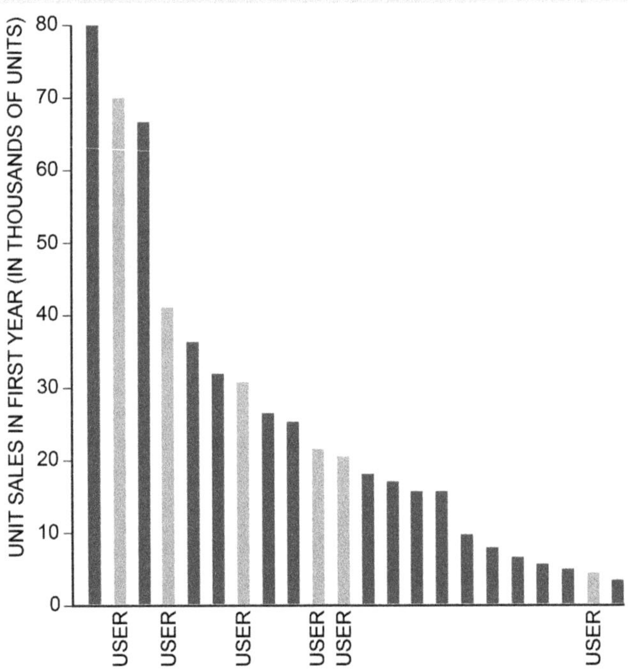

Figure 4: Five of six user-generated products in the top third of all launched products (USER: user-generated product, no label: nonuser-generated product). (Adapted with permission from Nishikawa *et al.* (2013). Copyright © 2012 Elsevier B.V.)

Source: Nishikawa *et al.* (2013).

Idea evaluation means that its technical feasibility, market attractiveness, and its strategic fit are assessed. Based on the evaluation results, some ideas are sifted out, while others move to the next stage for further refinement. Customer integration into idea evaluation focuses mainly on gathering customer feedback on ideas to help assess their market attractiveness. Many companies do not directly involve customers in this phase but use their gut feeling to judge how customers are likely to evaluate these ideas. Such approaches can, however, be harmful, as managers tend to harbor biases in their predictions of consumer preferences (Hattula *et al.*, 2015). Companies that involve customers tend to use qualitative customer integration methods to close the solution space by means of idea evaluation. On the other hand, quantitative idea screening tools are viable and effective if several ideas need to be evaluated and the ideas do not need elaborate explanations to understand them.

In the concept generation and product development phases, NPD teams develop ideas further to product concepts, functional prototypes, and final products. Customers can again contribute to opening (e.g., via lead user workshops) and closing tasks (e.g., via quantitative product testing). While customers' input in the early stages can largely influence the future course of an NPD, customer integration into the later stages is limited to incremental improvement suggestions and to market forecast estimations in the later stages. Besides its contribution to the innovation outcome *per se*, customer integration — notably of reference customers, such as opinion leaders on B2C markets, or B2B customers, who serve as role models in their industry — into this stage can also serve promotional purposes.

1.4 Customer integration for service innovation

Services differ from manufactured goods in many ways, but the two characteristics that most influence innovation management are their intangibility and the interaction between production and consumption. The intangibility of most services makes differentiation more difficult, because this makes identifying and controlling attributes harder. The near simultaneous production and consumption of many service offerings blur the distinction between process (how?) and product (what?) innovation and demand the integration of back- and front-end operations.

For example, in our study of 108 service firms in the UK and USA (Tidd and Hull, 2003, 2006), we found that a strategy of rapid, reiterative redevelopment (RRR) was associated with higher levels of new service development success and higher service quality. This approach to new service development combines many of the polar extremes of radical and incremental innovation's benefits but at lower costs and with fewer risks. This strategy is less disruptive to internal functional relationships than infrequent, but more radical, service innovations. In addition, the strategy encourages knowledge reuse through the accumulation of numerous incremental innovations. For example, the American Express Travel Service Group implemented an RRR strategy. In the previous decade, the Group had only introduced two new service products. The function of vice-president of product development was created, cross-functional teams were established, a formal development process was adopted, and computer tools, including prototyping and simulation, were deployed. In the interim, the group has developed and launched more than 80 new service offerings and has become the market leader.

Innovation in services is far more than the mere application of information technology (IT). In fact, IT investments' disappointing returns in services has led to a widespread debate about the causes of and potential solutions for the "productivity paradox" in services. Service innovations, which make significant differences to the ways customers use and perceive the service provided, require service providers to make major investments in process innovation and technology, as well as in skills and working methods to change the business model and make major marketing changes. Estimates vary, but returns on investment in IT alone are around 15%, with a typical lag of 2–3 y when productivity falls, which it often does, but when combined with changes in an organization and the management, these returns increase to around 25%.

Most research and management recommendations are based on manufacturing and high-technology sector experiences. People mostly simply assume that such recommendations apply equally to managing innovation in services, although certain aspects differ fundamentally. There is a clear need to distinguish what, if any, of what we know about managing innovation in manufacturing is applicable to services, what knowledge should be adapted, and what is specific and different.

Generic good practices do exist and apply to the development of both manufactured offerings and service offerings. However, these practices must be adapted to different contexts, specifically to the scale, complexity, and degree of the offerings' customization, as well as to the uncertainty of the technological and market environments. It is critical to match the configuration of development's management and organization to the specific technology and market environment.

The service sector includes a very wide range and a great diversity of different activities and businesses, ranging from individual consultants and shopkeepers to huge multinational finance firms and critical nonprofit public and third-sector organizations such as government, health, and education. Consequently, great care needs to be taken when generalizing about service sectors. Nevertheless, these service organizations share certain characteristics that set them apart from typical manufacturing operations: tangibility, simultaneity, storage, customer contact, and location.

First, goods tend to be tangible, whereas services are mostly intangible, even though you can usually see or feel the results. Consequently, performance and quality perceptions are more important in services, particularly regarding the difference between customers' expectations of a service and the perceived performance. Customers are likely to regard a service as good if it exceeds their expectations. Perceptions of service quality are affected by the following:

- tangible aspects — the appearance of the facilities, equipment, and staff
- responsiveness — prompt service and willingness to help
- competence — the ability to perform a service dependably
- assurance — the knowledge and courtesy of the staff and their ability to convey trust and confidence
- empathy — providing caring and individual attention.

Second, simultaneity describes the missing lag between the production and the consumption of services. Most goods are produced well in advance of consumption to allow for distribution, storage, and sales. In contrast, many services are produced and almost immediately consumed. This creates problems with quality management and capacity planning. It is harder to identify or correct errors in services and more difficult to match supply and demand.

The third characteristic is storage: Normally, services, for example, providing a seat on an airline, cannot be stored, although some services, such as providing utilities, have a potential for storage. The inability to own stocks in services can create problems regarding matching the supply and demand, that is, the capacity management. These problems can be dealt with in a number of ways: Pricing can help smooth fluctuations in demand by, for example, providing discounts at off-peak times. Where possible, employing part-time workers, or outsourcing, can provide additional capacity at peak times. In the worst cases, customers can simply be forced to wait for the services by queuing (Kotler and Keller, 2006).

The customer contact of the operations is another characteristic that defines the difference between goods and service. Most customers have little or no contact with the operations producing goods. Conversely, many services demand high levels of contact between the operations and the ultimate customer, although the level and timing of such contact vary. For example, medical treatment may require constant or frequent contact, but financial services require only sporadic contact.

Owing to the contact with customers and the near simultaneous production and consumption of services, the location of service operations is often more important for services than for operations producing goods, and the final distinguishing characteristic. For example, restaurants, retail operations, and entertainment services all favor proximity to customers. Conversely, manufactured goods are often produced and consumed in very different locations. Consequently, the markets for manufactured goods also tend to be more competitive and global, whereas many personal and business services are local and less competitive. For example, only around 10% of services in advanced economies are traded internationally.

These five service characteristics have to be taken into account when designing and managing new service development. Certain research findings regarding NPD will have to be adapted, or may not apply at all. In addition, the diversity of service operations means we need to tailor the organization and management of service development to different types of service contexts.

In practice, most operations produce some combination of goods and services. It is possible to position any operation on a spectrum ranging from "pure" products or goods to "pure" services. For example, a restaurant and a retail operation both have tangible goods on offer, but the service provided is mostly equally important. Conversely, most manufacturers now also offer customers some after-sales service and support.

However, the distinction between goods and services remains important, because the differences in their characteristics demand a different approach to their management and organization. As illustrated in Figure 5, it is perhaps better to think of any business or operation as offering a bundle of benefits, some of which will be tangible, some not. One can decide on the most appropriate mix of products and services to produce by taking these benefits into consideration (Hull and Storey, 2016).

Those businesses with the highest level of new service content tend to avoid overcomplicating their customer base. They are usually firms with fewer key customer segments accounting for a high proportion of their total revenue. Such firms integrate these specific customers into their innovation process in order to understand their needs and to streamline their service innovation accordingly. This suggests that customer complexity can be a barrier to effective innovation in service businesses. The rise of "no frills" air services in the USA and Europe since the mid-1990s,

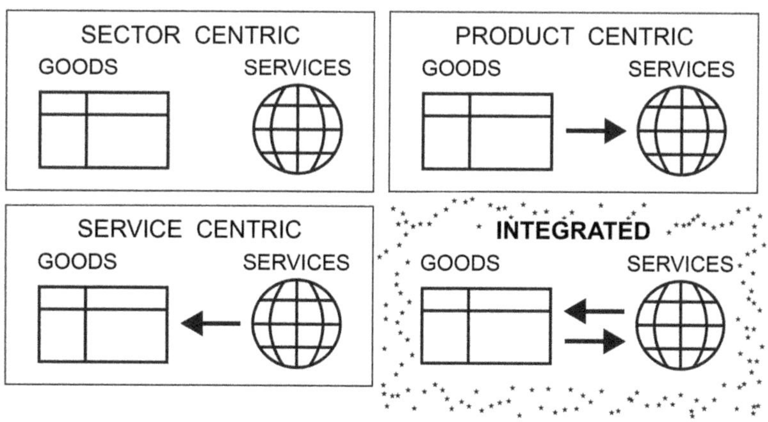

Figure 5: Integrating physical good (tangible) and service (intangible) innovation (Hull and Storey, 2016).

such as Southwest, Ryanair, and EasyJet, is typical of a focus on service strategy. It seems that a focus on the procurement and service delivery process also supports stronger. High innovators tend to focus their purchases on fewer larger suppliers and are less vertically integrated. They therefore focus on fewer internal processes within the overall value chain.

Research note: Types of service organization for innovation

Studying more than 100 service organizations, Tidd and Hull (2003, 2006) identified four distinct patterns or configurations offering different advantages.

Client project orientated. This type of innovating configuration follows structured processes, such as quality function deployment (QFD), to identify and influence customer requirements. These processes are mapped and continuously improved. The voice of the organization's customers and their early involvement in need fulfillment are integrated into this configuration. This configuration is strong in respect of organization but weaker in respect of tools/technology, such as the technological sophistication of either its knowledge or its IT. Nevertheless, the art and craft of project management, which is somewhat analogous to batch production in goods industries, provide a strong, yet flexible, type of enabling control over the development and delivery of customer-focused services. This configuration can achieve high levels of service delivery, service delivery on time to market, and cost reduction.

Mechanistic customization. The type of innovating configuration involves external customers in its product development and delivery process decisions by means of electronic interchanges. Standardization is a key factor in controlling the relationship, while electronic links are used to exchange data with its customers and suppliers. A key process control method is to set standards for projects and products, with customers helping to set these standards to meet their requirements. The configuration's electronic interchanges with its customers allow it to routinely adapt its service innovations to the market demand.

Hybrid knowledge sharing. In this type of configuration, the innovating company integrates the customer less directly. Employees are cross-trained, corewarded, and organized in groups, which reinforces their team

(*Continued*)

(*Continued*)

identity. Group systems are usually rather self-contained, which may be one reason why organizations with this configuration are more likely to value knowledge, reuse it, and share it to achieve a balanced portfolio of performance advantages. This configuration is strong in organization, tools, and system integration but lacks formal processes. Such a self-contained group of people become experts in developing and delivering products. Consequently, this type of configuration provides some of the advantages of codified knowledge but with bureaucratic forms exerting far less hierarchical control.

Integrated innovative. Colocated, cross-functional teams with a flat hierarchy characterize the integrated innovative organization's configuration. Regardless of rank, communications are open, whether face to face or via email. This configuration's technical base utilizes expert systems and management information systems. Market-oriented team members provide the customer perspective and work directly with the R&D team members to integrate the customer perspective into the innovation. Responsibility for the work is shared, and partnering is practiced throughout the value chain. Cross-functional teams and physical colocation facilitate dense communication about and collaboration on new service innovations.

None of these different service organizations is optimal in all contexts, but different organizational configurations perform best in different cases or contingencies. The integrated innovative configuration is the most innovative, the mechanistic customization one the most cost efficient, the hybrid knowledge sharing one the best for overall performance, and the client-project-orientated configuration the best for service delivery.

Source: Reproduced with permission from Tidd and Hull (2003) (Copyright © 2003 Imperial College Press) and Tidd and Hull (2006) (Copyright © 2006 John Wiley and Sons).

All four configurations in the research note "Types of service organization for innovation" have one or more significant effects on performance. Each appears to have evolved, or acquired, sufficient good practices to at least be viable in niche markets. The client-project-orientated configuration reduces the time to market and improves service delivery by focusing on customer requirements and project management. The mechanistic customization configuration reduces costs by setting standards and by involving the suppliers and customers. The hybrid knowledge sharing configuration provides a combination of innovation and efficiency by promoting team work and knowledge sharing. The integrated innovative configuration increases innovation and quality by means of cross-functional groups supported by groupware and other tools and technology, although the increased coordination increases the service development's time and costs.

Examination of the actual measures suggests that each of the four organizational configurations have several common elements, including

- the organizational mode of bringing people together
- the control mechanisms, which are either impersonal (standards, documentation, common software) or interpersonal (collocated teams)
- the shared knowledge and/or technical information base
- the external linkages to customers and other partners (e.g., suppliers).

The ambition to involve customers in the technical and market information is due to the desire to contribute relevant knowledge to the organization, which will allow the innovating entity to reduce the market uncertainties and the technological uncertainties associated with NSD. Cross-functional teams encourage collaboration and information sharing, thus improving the gathering, distribution, and utilization of such knowledge.

The model in Figure 6 describes the NSD process as a cycle and stresses the importance of teams, design tools, and organizational culture as enablers of service innovation throughout the NSD cycle planning and the execution progress, encompassing the design and analyses, as well as the development and full launch. Afterwards, the refinement and planning of new service innovations start, indicating an iterative, cyclical approach rather than a linear planning one (Tax and Stuart, 1997; Johnson *et al.*, 2000).

Similar to customer integration throughout NPD, service organizations can integrate customers into all four phases of the NSD process by means of activities that open or close the solution space. In the service

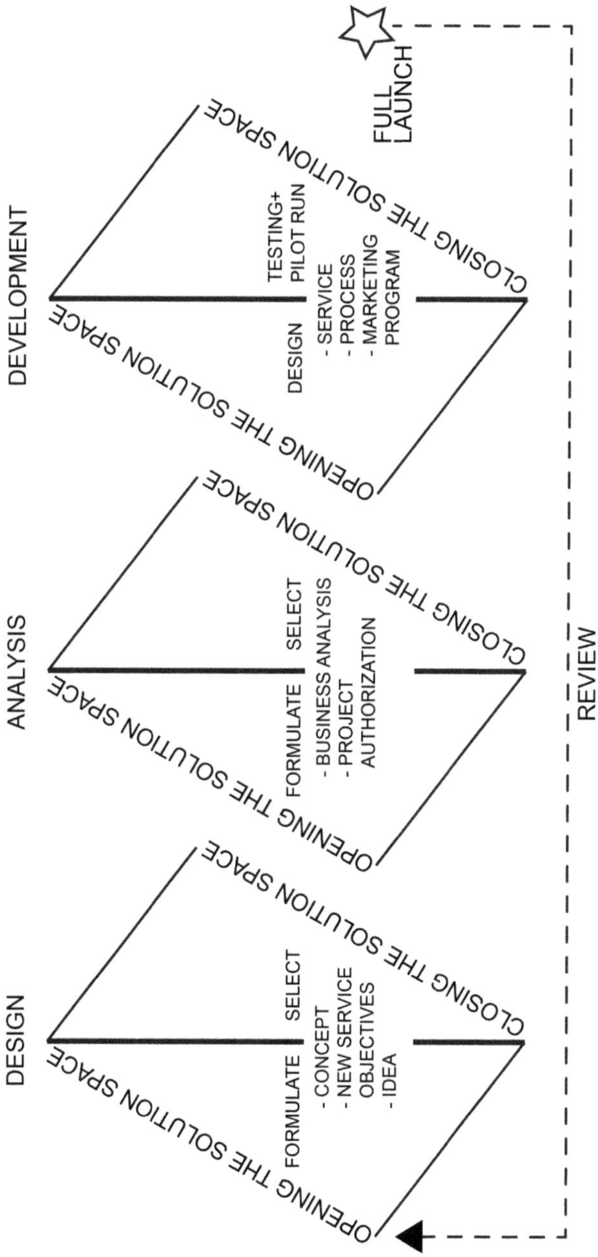

Figure 6: Customer integration in NSD.

design phase, customers can contribute to opening the solution space by helping discover latent needs as a basis for a new service development strategy in indirect customer integration approaches (e.g., latent needs identification through ethnography), or by inspiring new service idea generation in direct customer integration approaches (e.g., idea-generation workshops to jointly develop initial service blue prints). Customers can close the solution space by providing feedback on service blueprints and by comparing and participating in mock service delivery processes to provide insights into service attractiveness and the adoption likelihood, as well as feedback for service improvement.

The central argument for integrating customers into activities to open and close the solution space in new service creation is to provide superior customer experiences through a higher fit with customer needs and by drawing on customers' creative potential. In addition, customer integration in NSD can reduce the time to market and the NSD cost.

Customer integration in action: DOME

The Designing Out Medical Error (DOME) project used design-thinking methods to change medical services and clinical practices in order to improve patient safety. Design methods, ethnography, and process mapping were used to identify problems and develop solutions.

Rather than listening to what the managers, medics, and patients reported, the DOME researchers followed good ethnographic practice by shadowing the medical staff and patients to identify how the current services were provided. These were explored and challenged by using focus groups of staff and patients, as well as by applying design methods to develop prototype solutions that could be tested in real work settings.

For example, to help encourage good hygiene practices and reduce the spread of infection, a single, mobile, free-standing Care Station was developed, tested, and implemented. This Care Station included hand gel, gloves, aprons, a drug locker, waste and needle bins, a chart surface, and storage slot.

Source: http://www.innovation-portal.info/wp-content/uploads/DOME-final-version.pdf (accessed 30 August 2017).

1.5 Types of customers

Generally speaking, customers include all individuals who potentially benefit from a new product by using or consuming it as soon as it appears on the market (von Hippel, 2005; Piller and Walcher, 2006). This definition of customers includes current customers, current users, or potential customers. Figure 7 describes these different customer types schematically. Current customers are customers who have bought at least once from the innovating company in question. Current users include not only current customers but also customers who use or consume competitors' products in an innovation's product category or in a related product category. Current users comprise the company's current customers and customers of competitors. Potential customers are all individuals who fall into a new product's target group and may thus comprise current customers and noncustomers, as well as current users and nonusers. Users with domain-specific skills, such as lead users, who are understood as users who experience needs in a new and important area far earlier than the majority of customers and thus profit decisively from having this need satisfied, form another customer type (von Hippel, 1986).

The customers chosen for integration into NPD/NSD are pivotal. Current customers are not necessarily attractive future customers and

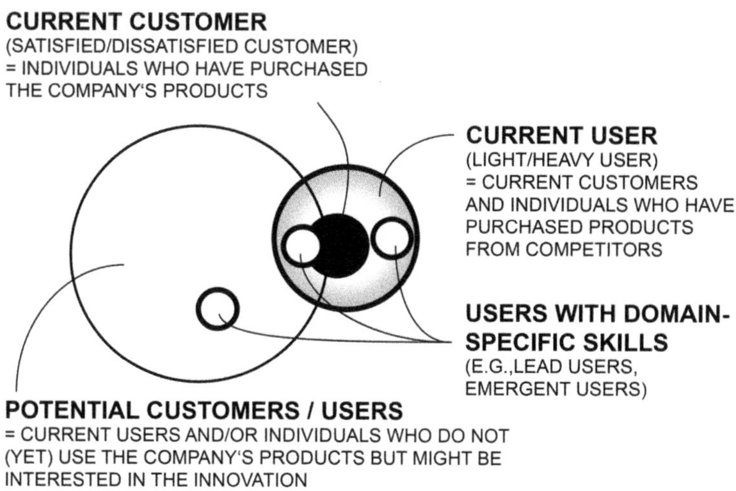

Figure 7: Categorization of customer types.

integrating only current customers may be misleading. In a seminal paper, Christensen and Bower (1996) describe how the largest manufacturers of 5.25-inch disk drives applied customer integration wrongly for the new 3.5-inch drives by only focusing on current customers. In the 1980s, they tested the market appeal of the new drives by showing leading customers of the existing product line the prototypes and asking them to evaluate the new drives. These customers, established desktop computer manufacturers, showed little interest in the new drive, because it did not address their need for capacity and performance. Consequently, corporate sales forecasts and expected profit margins of the 3.5-inch drive were pessimistic, leading to the abandonment of the 3.5-inch drives development of the manufacturers who led the 5.25-inch market and their refocus on improving the 5.25-inch drives, incorporating better heads and developing new recording codes. The customer integration's strong focus on current customers prevented the study of potential new customers' needs. Customers in the emerging laptop market were not considered, thus allowing the competition to step in, cater for, and grow with this market. Learning curves in the laptop market helped the 3.5-inch drive manufacturers improve their product and finally replace the 5.24-inch drives in the desktop markets throwing the former top bulls in a struggle for survival.

Current customer integration is also disadvantageous if an innovation destroys current customers' competences and when an innovation requires customers to unlearn specific behavior. In these cases, current customers need to invest in certain competences and resources that do not complement but substitute their prior investment in time, money, and effort (Henderson and Clark, 1990; Afuah, 2000). For example, a new keyboard with a different key arrangement to improve typing speed requires customers to unlearn old typing behavior and relearn a new way of typing. In such cases, current customers are reluctant to adopt the innovation and are unlikely to provide meaningful insights if integrated into NPD/NSD (McDermott and O'Connor, 2002; Garcia *et al.*, 2007).

Integrating existing customers into mainstream markets is, however, not doomed *per se*. On the contrary, in many cases, integrating and satisfying current customers pay off, as it is generally easier to retain existing customers than to win new ones. Insights into current customers' needs and wants are beneficial for the development of incrementally new products along continuous paths. Current customer integration also has its merits for technologically radical products that target mainstream markets. Understanding current customers profoundly and trying to better satisfy their needs lead companies to explore new technologies to produce superior benefits. Large investments in such new technologies with relatively uncertain returns might be easier to justify to shareholders of the results of concept tests with current customers are promising (Govindarajan *et al.*, 2011).

Beside the differentiation between existing and potential customers, the differentiation between ordinary users, who form the majority of current and potential customers, and users with domain-specific skills is relevant. Domain-specific skills refer to the knowledge, know-how, and abilities required to carry out a specific task. Individuals with high skills in a specific domain perform tasks in this domain better than those with low domain-specific skills.

Research note: Domain-specific skills

The theory of domain-specific skills dates back to the novice–expert comparisons of Simon, Ericsson, and colleagues (Simon and Chase, 1973; Ericsson and Lehmann, 1996), who studied differences in the utilization and recall of knowledge between experts and beginners. In their studies of chess players, Simon and Chase (Chase and Simon, 1973; Simon and Chase, 1973) found that experts recall the position of figures on a chess board better than novices. After looking at the chess board for 5 seconds, beginners can, on average, recall only four, while experts recall more than 20 figures correctly. Proficient players are better at memorizing domain-specific new knowledge and access it quicker and more correctly than novices. Not only the quantity of knowledge but also its structure differs between novices and experts.

Amabile (1983, 1996) includes domain-specific skills in her componential theory of creativity as one of three key elements of creative activity (along with creative thinking, skill and task motivation). Creative ideas often represent combinations of previously unrelated areas and thus require individuals to master and to have cognitive access to specific domains in order to act creatively. This view extends domain-specific skills from

(*Continued*)

(*Continued*)

cognitive skills to creative skills and understands domain-specific skills as a precondition for domain-specific creativity.

Domain-specific skills can, however, backfire if solutions are not within the specific domain in which a person has a high skill level. Skills in a certain area consciously and unconsciously steer thinking and lead the search for solutions to any problem in this area of expertise. If the solution to a problem resides outside their area of expertise, individuals with high levels of domain-specific skills will nevertheless look for solutions within this domain, with negative consequences for their ability to develop a solution. The time invested in the development of skills in one domain is not available for the development of skills in other domains. The higher the expertise in this area, the stronger the bias.

Users with domain-specific skills develop new product ideas that correspond to this domain. Give such users the same NDP task, and they will produce different ideas, which are based on their domain-specific skills. Users with high technical skills produce ideas that are technically more feasible; those who are trend savvy and informed about the latest technical gadgets produce more technically novel products but less feasible ones. Users with high sustainability knowledge produce product ideas that have a more positive environmental and social impact (Kristensson and Magnusson, 2010; Schweitzer *et al.*, 2014). The choice of customers influences the customer integration results. Consequently, the goal of a customer integration approach should be scrutinized and the most appropriate customers should be selected.

1.6 The strategic cornerstones of customer integration

1.6.1 *Customer orientation*

Customer orientation is a strategic focus on satisfying customers and valuing their insights in appreciation of the value of their satisfaction and their insights as a means of attaining one's business objectives (Schmookler, 1966). Originally, innovation research and practice coined the term customer orientation, or market pull, as a juxtaposition of technology orientation, which is also understood as technology push. The advocates of the one approach disagreed fiercely with the other camp regarding these orientations' values. Technology push follows the exploration of technology's potential as the initiating element for innovation and finding a way to use this technology to satisfy customer needs in superior ways. Customer orientation or market pull starts with understanding customer needs as a basis for choosing the technology to meet these needs.

In recent years, the positions of the two approaches have moved toward an integrative perspective due to the need to simultaneously have a technology and customer orientation in NPD/NSD (Brem and Voigt, 2009). In industrial practice, the emergent use of technology roadmaps to identify long-term developments in markets and technologies and to link planned technology, market trends, and product/service developments is an obvious sign of these merging perspectives.

Simultaneously, the customer integration concept has evolved. Narver *et al.* (2004) developed the responsive and proactive market orientation constructs to distinguish between two customer orientation approaches. Responsive market orientation corresponds to a proclivity toward customers' expressed immediate needs (Day and Shoemaker, 2005), which is often associated with the deleterious effect of the served market's tyranny (Hamel and Prahalad, 1994). Proactive market orientation captures a set of behaviors associated with detecting and satisfying latent customer needs, embracing future markets. Future market orientation is a focus on the value and potentials of new markets, analyzes existing markets' latent and future needs, and represents a willingness to cannibalize existing products, services, and markets for own innovations (Chandy and Tellis, 1998). Different researchers have repeatedly demonstrated future market orientation's positive impact on radical innovation and that of current market orientation on incremental innovation (Jaworski *et al.*, 2000; Lindgren and O'Connor, 2011; Narver and Slater, 1990; Slater and Narver, 1999). These findings correspond to other innovation studies that confirm a positive correlation between proactive information search and corporate innovativeness (Kitchell, 1995). A proactive disposition toward meeting potential future customers' latent needs also seems consistent with a long-term investment in a company's future growth, rather than myopically striving to maximize current profits. This disposition encourages relinquishing quick short-term wins for long-term superior customer value (Pierce and Delbecq, 1977; Mullins and Sutherland, 1998; McDermott and O'Connor, 2002), resulting in technical and organizational innovativeness rather than immediate market performance effects (Han *et al.*, 1998).

A responsive customer orientation is also related to existing performance parameters. This orientation focuses NPD/NSD teams too narrowly on their established customers' requirements, thus obscuring the view of the changes in customer needs over time. This obscured view prevents companies from detecting radical innovations with a disruptive nature that established customers do not, at first, find attractive. Such innovations are high in market innovativeness but not necessarily so in technological innovativeness. Newcomers may provide an inferior technology as a cheaper alternative to mainstream offers and first enter niche markets that eventually raid the incumbent's customer base (Christensen and Bower, 1996; Govindarajan and Kopalle, 2006). For example, high-quality camera producers, like Leica, did not invest in digital camera technology, because their customers (the high-end consumer market and professional photographers) focused strongly on the quality of their photographs, which digital cameras lacked for a long time. However, digital camera technology developed swiftly and Leica struggled to catch up with its customers, who had first ignored and later embraced digital cameras.

Further, responsive customer orientation provokes knowledge exploitation, which refers to the improvement and refining of existing knowledge. Knowledge exploitation results in adaptive learning, a customer-led learning approach to learning about, and satisfying, customer exigencies and does not contribute to radical innovation (Baker and Sinkula, 2005, 2007). Proactive customer orientation fuels knowledge exploration, pushing companies to acquire entirely new competences and to generate new insights and skills from their customer-orientation experiences. This orientation involves allocating financial and human resources to explorative search and obtaining distant knowledge from new types of customers (Danneels, 2003). Knowledge exploration entails generative learning, which leads companies to move beyond the current customer boundaries, thus challenging existing assumptions about their markets and fueling searches for unique insights and inspirations (Kitchell, 1995; Amabile *et al.*, 1996). Consequently, a proactive, explorative approach might be a more promising orientation for successful innovation, although it also has its downsides. Explorative learning can only take place when a company has the required absorptive capacity to incorporate and utilize the information (Bindroo *et al.*, 2012; Deshpandé *et al.*, 1993). Companies might be unable to absorb too distant market knowledge (Lane *et al.*, 2006; Maes and Sels, 2014). Further, exploration can only occur if the company has the absorptive capacity to identify, assimilate, and utilize this knowledge (Cohen and Levinthal, 1990). Since prior knowledge is a prerequisite for absorbing new knowledge (Roberts *et al.*, 2012), companies experience difficulties when future markets are unfamiliar and unrelated to the existing ones. These difficulties include the lack of an internal organizational structure to help companies fully seize market opportunities and a lack of knowledge about business models that help leverage the market's potential returns (McDermott and O'Connor, 2002). Consequently, some studies draw attention to the potentially curvilinear relations between knowledge exploration and radical innovation (Bao *et al.*, 2012). Others allude to the need for knowledge exploitation as a prerequisite for successful exploration, because acquired explorative customer knowledge needs conversion into commercial knowledge through exploitation (Lane *et al.*, 2006; Maes and Sels, 2014). In addition, antagonizing current customers does not sound like a sustainable business approach, given that it is easier to retain existing customers than to win new ones. Consequently, balancing between proactive and responsive orientations on a corporate level is sensible, while different NPD/NSD projects emphasize reactive or proactive orientation, depending on whether the outcome is expected to be incrementally or radically new.

1.6.2 *Manufacturer active paradigm (MAP) and customer active paradigm (CAP)*

The manufacturer active paradigm (MAP) and the customer active paradigm (CAP) are two different ways of dividing innovation tasks between manufacturer and customer. Figure 8 illustrates the key difference between MAP and CAP. MAP is the traditional customer integration into

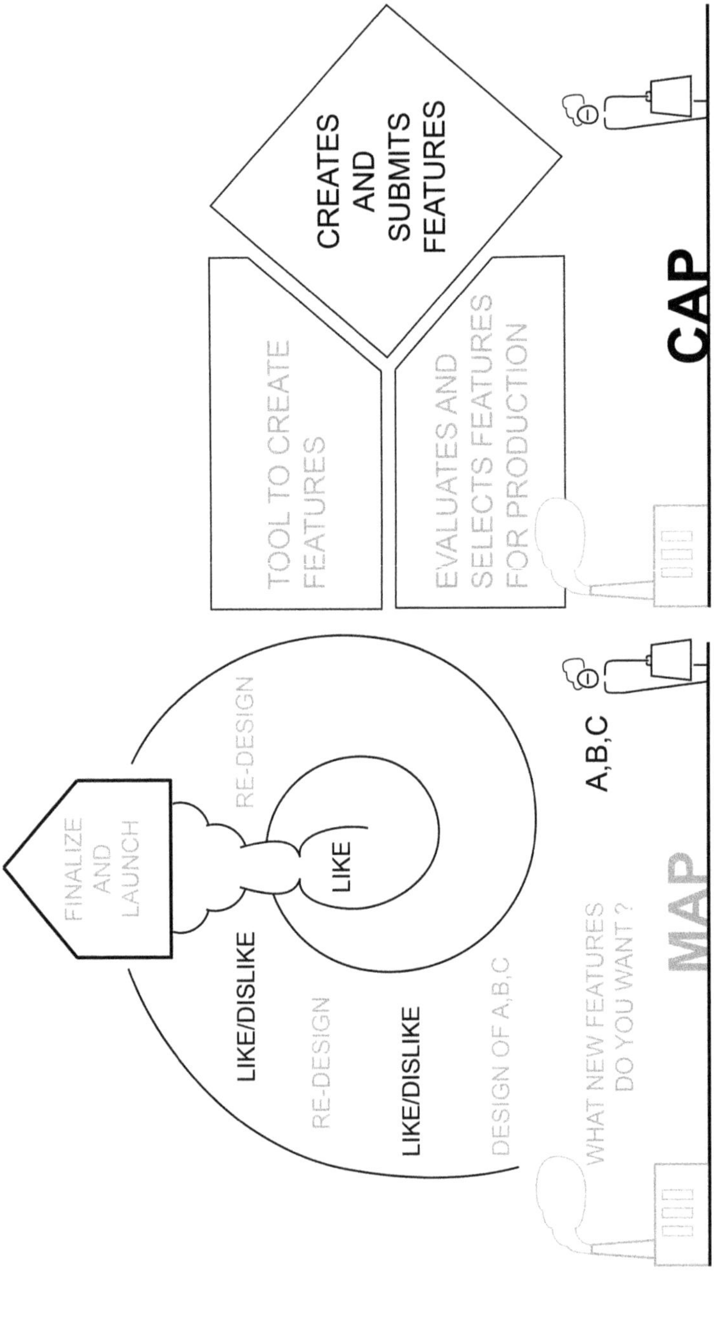

Figure 8: MAP and CAP.

NPD/NSD that a lot of firms have practiced for many years. This process concentrates on customers as providers of needs information. Manufacturers following MAP carry out new product or service development by iteratively asking customers about their needs and the fulfillment of these through specific offerings. Innovation starts with asking customers about unfulfilled needs. This first step also includes customer integration's indirect approaches to detect latent needs. Thereafter, the company internally creates concepts that help satisfy these needs before presenting the concepts to their customers to evaluate the degree to which these satisfy their needs. The company next refines the concepts into products, services, or product–service combinations and again presents them to their customers. At the later stages, customer integration thus focuses on methods by which the solution space can be closed. Applying this approach, the company indirectly integrates its customer by first retrieving information on their needs from attractive customer segments, then developing an offer that is aimed at meeting their needs, and testing its acceptance throughout the innovation process via concept tests, prototype tests, product tests, and market tests. In the course of the NPD/NSD process, the manufacturer collects and analyzes customer data in order to develop and refine the product or service concept in iterative steps to finally meet customer requirements in the best way possible. The role of customers is limited to conveying their needs and providing feedback on their satisfaction when asked to do so.

CAP puts the customer center stage. The customer acts as a provider of solution information and thus a generator of new product and service ideas and concepts. Direct methods to integrate customers into tasks to open the solution space correspond to this approach. This approach promotes the central idea that CAP helps transfer tacit and sticky usage-experience knowledge from the customer to the manufacturer. Information stickiness refers to the "incremental expenditure required to transfer that unit of information to a specified locus in a form usable by a given information seeker. When this cost is low, information stickiness is low; when it is high, stickiness is high (von Hippel, 1994, p. 430)." What people like or dislike in T-shirt designs may be difficult to articulate. When the innovating company provides tools with which the customers can easily sketch what they like, they might find it easier to show directly what they like rather than talking about what they like. The innovating company can then use this design input to produce the T-shirt.

If customer involvement by means of this approach does not end with idea generation but extends to the development of functional prototypes, the manufacturer's task in the innovation process is reduced to discovering and selecting those offers worth realizing, preparing the offering for mass production, realizing it, and marketing it. Proponents of the customer active paradigm criticize MAP for its resource intensity and highlight the efficiency that CAP gains.

For example, having internal designers create a T-shirt design involves several iterative processes between the T-shirt manufacturer and the customers. First, the innovating company needs to ask the customers what designs they would like to see on a T-shirt, and then the company draws different designs based on their descriptions. Later it again presents the designs to them. If the product category is not T-shirts but complex industrial goods, this step can take a while and, on returning to the customers, the innovating company may find that it misunderstood what they wanted or the customers might during the lengthy process have changed to another company's offerings to satisfy their needs. In the consumer goods industry, the companies Threadless and Spreadshirt demonstrate that CAP facilitates the transaction process dramatically. Both companies allow their customers to create and upload their designs via information and communication technology (ICT) tools.

It is difficult to judge which approach is better than the other. The risk of misunderstanding the customer-need information incorporated in MAP can be reduced through the accurate design and execution of the customer integration activities, but CAP is not the perfect alternative. CAP risks potential quarrels with customers about intellectual property (IP) rights and the manufacturer not reaping the innovation profits, because his value-creating contribution is less. If users start producing and marketing these innovations, CAP can even turn former customers into competitors. These risks are more prevalent when manufacturers integrate corporate customers (B2B) than when they integrate consumers (B2C) (Table 1).

Table 1: Strengths and challenges of MAP and CAP.

MAP	CAP
Strengths:	Strengths:
• Manufacturer as the central driver of innovation	• Low cost of transferring sticky customer information
• Less risk of problems with IP rights	• Innovation is less resource intensive
Challenges:	Challenges:
• High cost of transferring sticky customer information	• Risk of quarrels with customers about IP rights
• Resource-intensive innovation tasks	• Relinquishing part of the revenue to customers
	• Today's customers turn into tomorrow's competitors

Customer integration in action: The rise and fall of Quirky

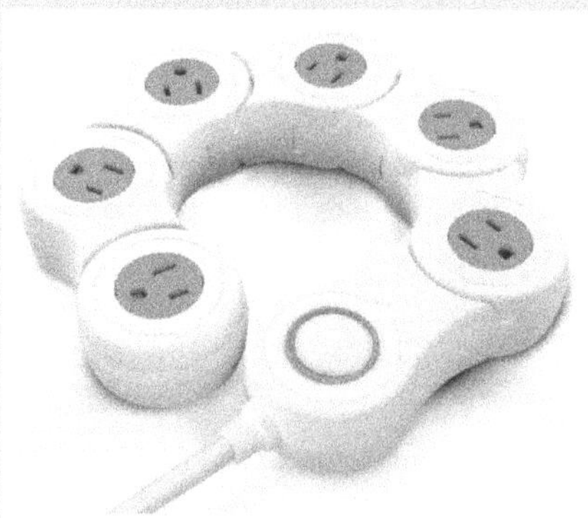

Figure 9: Quirky's bestselling item (more than 490,000 sold) — a bendable power strip called Pivot Power.

Source: http://www.business-insider.com/simple-successful-inventions-quirky-ben-kaufman-2012-12?IR=T (accessed 30 August 2017).

Quirky, a US virtual cocreation platform, built its business model on the customer active paradigm. Registered users submitted product ideas and refined the ideas to arrive at the final prototypes. They also chose the final prototype. Quirky management selected a set of the highest rated ideas that idea submitters presented at a local public evaluation event. At this event, the Quirky management and the community members commented, discussed, and finally chose the products that are put forward to production (Figure 9).

In its idealized model, the role of the company was limited to finding a manufacturing partner for production, contracting retailers for distribution, and assuring a maximum period of 80 days from selection to delivery. Quirky divided the NPD tasks

(*Continued*)

(*Continued*)

into smaller subtasks to allow a large number of people with different skills to participate in the cocreation. The type of tasks included submitting a rough idea, providing a product name, or a final design. Depending on the specific cocreation task, these cocreators received a certain share of the profits, and all the contributors were mentioned on the product's package. Quirky's core customer promise was to launch innovative new products every week. This approach allowed Quirky to produce highly novel new products that clearly addressed important customer needs. Ben Kaufman founded Quirky in 2009, raised USD 97 million in venture capital within the company's first 3 y, had rallied a community of over 300,000 users by 2012, and filed for bankruptcy with a debt of USD 100 million in 2015.

The prime reason for Quirky's difficulties lay in the decisions and refinements required after the concept had been refined. Quirky had to invest considerable internal resources to ensure the products achieved minimum quality levels and manufacturability. As a young and small company, the development of the required internal knowledge about the materials and technologies required, as well as the manufacturing and distribution possibilities of the products, which ranged over 26 product categories, could not keep pace with the speed at which the community churned out new product ideas. In 2011, the submitted new product ideas reached an average of 1,000 per week. The vast number of unrelated products sold in small quantities also made it impossible to achieve significant economies of scale or develop platform strategies. Many of Quirky's products received mediocre to poor online reviews.

Quirky did not follow CAP consequently and stringently from idea generation to launch but used an unhealthy mix of CAP and MAP. CAP switched to MAP as soon as the product concept design was ripe, with ideas pouring in and Quirky struggling to keep its promise to bring them to market, while also lacking the capacity to do so sustainably, as MAP required internal competences in the development stage that the company could not provide.

Sources: Marks and Hoyt (2013); Kornish and Ulrich (2014); Fixson and Marion (2016); and own web research.

Videos: *The path to Quirky*: https://www.youtube.com/watch?v=GqYeWg7igkU (accessed 30 August 2017), *Quirky's problems*: https://www.youtube.com/watch?v=fJ-8feQjXhQ (accessed 30 August 2017).

1.6.3 *The four basic strategic approaches of customer integration*

The strategic approach steers the way in which customer information is gathered and utilized throughout the innovation process. Based on the two types of customer orientation (responsive/proactive) and the two paradigms (MAP/CAP), there are four basic strategic customer integration approaches (see Figure 10):

- customer-led approach
- customer-focused approach
- customer-centered approach
- customer-driven approach.

Companies, business units or NPD/NSD teams following a customer-led approach focus on carrying out customer requests (Nijssen *et al.*, 2012). They have a responsive customer orientation and are based on the MAP. This approach concentrates on gathering needs information from existing customers. For example, imagine a manufacturer of chest freezers sold to supermarket chains to store chilled and frozen food. The manufacturer usually has only a few large customers, each with hundreds of stores throughout the country. Such a manufacturer would lose a considerable part of the company turnover if a single customer were to shop elsewhere. The

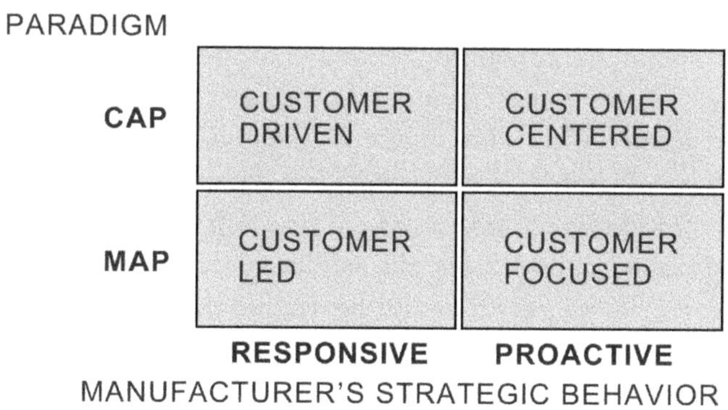

Figure 10: Four approaches to customer integration.

manufacturer is therefore eager to understand and fully satisfy each customer's needs. If a customer were to tell the manufacturer's sales representative that he needs a larger display depth — at least 620 mm instead of the regular 510 mm — and would buy at least 200 if they were offered, the salesperson would do his best to persuade his boss to provide a 620 mm solution. Sooner rather than later, the request would be presented to the R&D department's requirements specification. The R&D department might be flooded with similar demands, thus reluctantly putting it in the queue. Once realized, the salesperson might present the innovation to his customer, who might have forgotten about his request, changed his mind about the feature he needed, or decided that only 13 such fridges would suffice.

Customer-focused entities strive to create customer value with all their innovation activities (Homburg *et al.*, 1999; Gulati and Oldroyd, 2005). They follow a proactive MAP approach. Again, gathering needs information is focal, but the way this information is gathered and transformed internally differs substantially from the customer-led approach. The same manufacturer as above would differentiate between customer requests and latent needs. If a request is received, the company would first decide whether the idea should be realized and, if so, whether this customer request will be run as an individual adaptation process, or whether it is a latent need that a large part of the market would potentially find attractive. In the first case, requests are dealt with as a build-to-order business that would need orders. Manufacturers would also be clear about the customer having to pay for individualized offers in the build-to-order business and about the timeframe needed in the develop-to-order business. In the latter case, the requirement forms the basis of an innovation project and is used jointly with other customer insights as an inspiration rather than a strict specification. Wintersteiger is a company with three main pillars and the world market leader in field research equipment, as well as in service and rental solutions for ski and snowboard sport, and a leading manufacturer of advanced thin-cutting technology for lamellas production. This company institutionalized the customer-focused approach and has two different R&D divisions for its build-to-order business and for innovations that enter its standard product program. The staff and resources of these two units cannot be exchanged. This task separation helps Wintersteiger assure that build-to-order customization work does not deplete its innovation resources.

Companies following the CAP can also either take a responsive (i.e., customer-driven approach) or a proactive customer-orientation strategy (i.e., customer-centered approach).

In customer-driven entities, the customer is the key player, while the manufacturer is minimally involved in innovation (Desouza *et al.*, 2008). The manufacturer collects ideas and commercializes them rapidly or puts existing customer solutions together in product–service bundles. Customers create the solution individually or collectively, while the manufacturer's innovation task is very limited. This signifies that the manufacturer also releases control and acts as a mere facilitator by offering manufacturing capabilities.

Similar to the customer-driven approach, customers are involved not only as providers of needs information but also as providers of solution information in the form of ideas and concepts for new products, services, materials, or technologies. The company and its customers innovate together. Contrary to customer-driven innovation, the manufacturer does not hand over control to the customer but actively shapes the innovation outcome through its interaction with its customers and orchestrates the innovations that various customer communities have suggested.

Customer-driven approaches face several challenges. One challenge is that customers who cocreate ideas are not necessarily the target group for the created products and services. Ideally, customers should be the initiators and the users of new products and services; in reality, the two roles are rarely a perfect fit. In imperfect fit scenarios, initiating customers who become selectors might do a bad job due to the mismatch between their needs and those of the larger market.

If customers in such situations do have not only the creative power but also the power to choose, their choices might not mirror the target group's needs information. The danger thus lurks that their decisions might not lead to attractive solutions for the target group. Moreover, not all customers play their role in line with the manufacturer's expectations. For example, customers are reported to hijack idea competitions and evaluations to either let off steam or make fun of them. For example, the German fast-moving consumer goods company Henkel carried out an idea competition

Customer integration in action: Cocreating a car

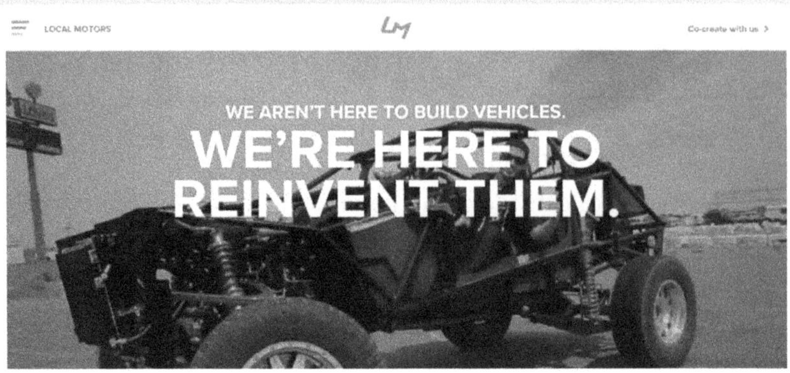

Jay Roger founded Local Motor in 2009 with the goal of having customers design and build their own cars. Nevertheless, he rather quickly realized that the individuals, who submitted and voted for car designs, were largely persons who had just finished design school and were looking for jobs in the design industry. With microfactory-produced cars' prices starting at USD 60,000, the target customers were financially well-off experiential buyers, who are interested in kit cars as a way to differentiate themselves from other car owners. Rogers allowed the crowd to vote for their favorite design and made sure the winner received a prize but clearly highlighted that he would personally choose the final design that would be realized.

for a creative, new bottle label for its dish-liquid brand Pril and allowed its customers to vote for their favorite design. Henkel planned to launch the two winning designs as limited editions. Many submissions featured inspiring colorful and flowery designs with filmy slogans. One submission, however, had brown color and a slogan saying that it would taste wonderfully of chicken. More than 33,000 labels were submitted and the chicken label won with 3,500 votes more than the second one. Henkel then used a jury of five experts to choose the final design for the limited edition and had to endure scathing comments from bloggers who felt that the company had not kept its promise.

Customer-centered organizations prevent such counterproductive energy from hampering their innovation outcome. They may allow customers to vote for their favorites but clarify from the beginning that the final decision resides with the company. They then study the customer requirements and needs behind the customer-suggested ideas or concepts and gauge their potential for the broader market before making their decision.

Another challenge of customer-driven approaches is especially relevant in B2B markets: If the customer carries out the majority of the innovation work, it is likely that he will also strive to reap most of the generated innovation value. Further, the customer might find it easy to switch to another supplier or to integrate backward and become a future competitor.

Customer integration in action: The launch of a new Nivea product line

Beiersdorf AG used Hyve as a consultant to run an online cocreation study in 2009. This study would involve customers in the generation and selection of new product ideas. To recruit participants, Beiersdorf ran conventional communication campaigns, as well as social media campaigns, involving opinion leaders and reputed bloggers, such as the "Undershirt Guy" (www.undershirtguy.com).

(*Continued*)

(*Continued*)

The cocreation experience used a gamification approach with interactive web 2.0 elements, such as a "developers' index," a list summing up all participants' contributions and providing contributions with an originality rating. Beiersdorf transported its customers into the imaginative role of product developers and asked them to evaluate 30 product ideas in terms of their innovativeness, market fit, added value, and other criteria. The company then calculated an index for each idea based on the customers' suggested evaluations. The participants commented on the ideas, provided improvement suggestions, and suggestions for specific usage situations. Each idea received more than 500 comments. The ideas and the comments inspired further development steps for "Invisible for Black & White." In the later steps, Beiersdorf focused its collaboration on experts from other industries (e.g., textile) and universities to achieve the right mix of chemicals for an antiperspirant that would prevent staining. Beiersdorf launched the product by partnering with distribution partners and promoters who would introduce the innovation to its key audience. In this way, Beiersdorf used CAP elements for innovation and also integrated partners beyond customers in a broader open innovation (OI) endeavor. Throughout the process, the company clearly remained the conductor of the innovation orchestra. This customer-centered development process led to Beiersdorf's most successful launch in its 130-y history. Within the first 9 months, Beiersdorf sold more antiperspirants than its major competitor sold in a year.

Source: Bilgram *et al.* (2013).

The tendency is for companies that follow customer-focused and customer-led approaches to be more successful with new market offerings than others, because proactive orientation allows companies to hold the reigns of innovation, while responsive orientation is reactive. Yet, the success of the different approaches ultimately depends on a company's specific competences. The key competences of successful customer-driven companies are usually production and logistics, while marketing and R&D play a minor role. A company that has very flexible production processes may be rather successful when following a customer-driven approach. In addition, companies that coordinate a focused value chain, source out innovation, and perhaps even manufacturing to others, but have excellent market access and offer distribution and communication as their key competences, may profit from customer-driven approaches. It is important to understand a company's approach to customer integration, because it influences how customer integration can drive an innovating company's business. The choice of customer integration method has to be compatible with this approach. In other words, using one or the other method that is incompatible with the current approach means that a company will need preparatory work to create a subunit comprising people with the appropriate customer integration approach. If it fails to do so, the method's execution is likely to lack effectiveness.

Chapter 2

Direct approaches to open the solution space: Users as creativity machines

2.1 Customers and creativity

Early creativity studies focused on extreme forms of creativity as manifested in eminent artistic and scientific performances (MacKinnon, 1961; Barron, 1968; Welsh, 1975). These studies shaped the understanding of creativity as a special trait residing only in a few outstanding minds. Creativity, however, also includes the more mundane trait of productive thinking and acting that solving everyday problems characterizes. This is the way productive thinking and acting generate content that is novel and useful (Stein, 1953; Lubart, 1994), which is also how customers undertake creative tasks in their daily life. They alter the form of a product or combine products in ways that contradict their typical applications, because they need to solve problems with the standard use of products or because a specific situation requires this. For example, consumers planning to attend a formal dinner might discover that their shoes need polishing but they have no shoe polish and the shops are already closed. A creative solution to the problem could include coloring the scuffs with magic marker or eye mascara and then buffing the shoes with baby oil or car wax.

When a company focuses on NPD/NSD, its customers provide creative input (1) by providing ideas in online idea competitions (Piller and

Walcher, 2006; Poetz and Schreier, 2012; Nishikawa *et al.*, 2013), (2) by sharing ideas in company-sponsored or independent online communities (von Krogh *et al.*, 2003; Bagozzi and Dholakia, 2006; Füller and Matzler, 2007; Fu *et al.*, 2008; Bayus, 2013), (3) by cocreating products, services (Lakhani and von Hippel, 2003; Bagozzi and Dholakia, 2006; Chatterji and Fabrizio, 2013), and their early concepts in, for example, lead user workshops (von Hippel, 1986; Olson and Bakke, 2001; Hienerth, 2007), or (4) by acting as user innovators who provide unsolicited ideas (Alexy *et al.*, 2011) or start entrepreneurial activities (Baldwin *et al.*, 2006; Shah and Tripsas, 2007; Hienerth and Lettl, 2011).

Customer integration in action: Bringing codesign to light

Zumtobel, an Austrian manufacturer of lamps and integral lighting solutions, involves innovation-affine customers demanding high quality NPD standards. The company differentiates between two customer types: architects and key accounts. Zumtobel uses an elaborated selection procedure to integrate appropriate architects. National company representatives ensure they are updated regarding the architectural world's developments and trends, know which architects are regarded as hip and innovative, and preselect candidates with a good chance of coming out on top in the project environment due to their reputation and prior business. Thereafter an employee from Zumtobel's strategic partner development group contacts the preselected architect for a first codesign project due to start 3–6 months after the initial contact. The goal is to create a nonstandard limited-edition lighting system that usually bears the name of the architect. These codesigned limited editions tend to be incrementally new with respect to light technology but radically new with respect to functionality and form. After 2 y, Zumtobel evaluates the limited-edition lighting system's success on the basis of its positive image effects and sales. The limited edition should capture 5–10% of the architect's lighting system volume. Only a few of these evaluated limited editions are further developed and standard products. The architects function as agents who need to assure that a certain number of their lighting systems are sold to their projects. The second type of customers is selected on the basis of their sales volume and includes retailers with many outlets, such as the supermarket chain Spar. Zumtobel codesigns lighting solutions with key accounts if they preorder a predefined number of units and if their reputation and size qualify them as reference customers for other clients.

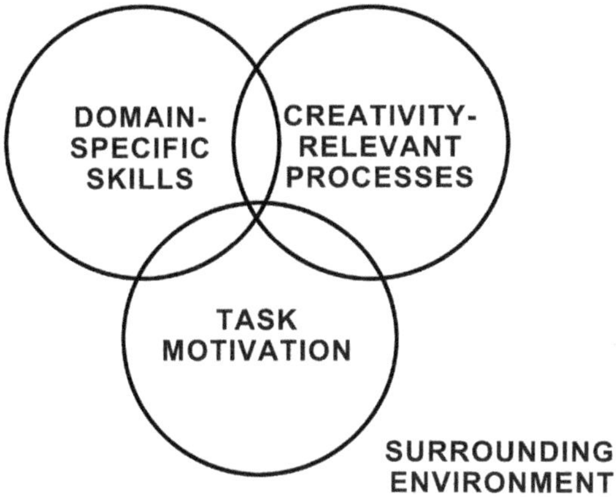

Figure 11: Amabile's three components of creativity.

In Amabile's (1983) three componential theory of creativity (Figure 11), an individual's domain-relevant skills (the skill set in a relevant domain from which an individual can draw), creativity-relevant processes (conducive cognitive and personality factors, such as risk taking, having new perspectives on problems, disciplined work style, idea-generation skills, tolerance for ambiguity, and self-discipline), and task motivation impact creative action. Beside the confluence of these three individual factors, the surrounding environment can curb or foster creativity. For example, an organizational culture that emphasizes the status quo, or top management's low-risk attitude, is counterproductive. Self-determination in carrying out tasks and a culture that regards mistakes as learning and improvement opportunities nurture creativity. According to this theory, creativity is highest when a person with high domain-specific skills, internal motivation, and creative-thinking skills acts in a creativity-supporting environment (Amabile, 1983, 1996).

In the context of customer integration into creative innovation tasks, this theory translates into a company having to understand (1) customers' capabilities, (2) their motivation, and (3) the nature of the problem prior to choosing and designing the right customer integration technique (Figure 12). In the first context, it is important to scrutinize the different capabilities that ordinary customers and expert users contribute.

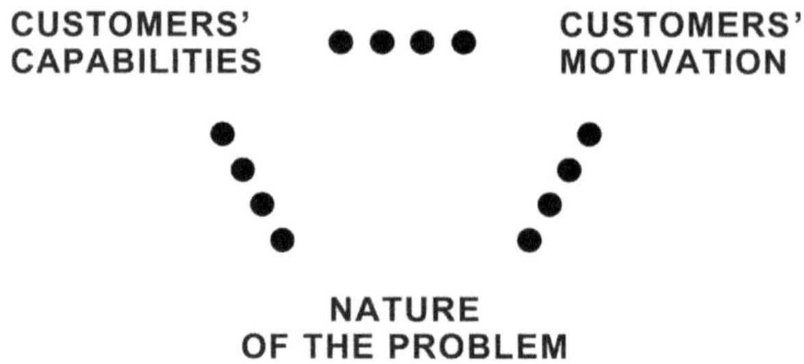

Figure 12: Factors influencing consumers' ability to contribute to creative innovation tasks.

Several studies show that ordinary users (i.e., lay persons who are current or potential customers) outperform R&D professionals in creative idea generation for new products, services, and product–service combinations. In the Bamed/MAM Group's idea contest, ordinary users generated ideas for improving the convenience of feeding babies with solid food. Corporate professionals also generated ideas. The CEO and the head of R&D, both blind to the ideas' origin, rated the ideas in terms of their originality, their ability to create customer benefit, and their feasibility. Ordinary users' ideas were more original and provided higher customer benefit, but were slightly less feasible than professionals' ideas (Poetz and Schreier, 2012). Other studies (Kristensson *et al.*, 2002; Magnusson, 2009) demonstrate that customers provide creative ideas for technical solutions in familiar contextual use scenarios (e.g., students generating ideas for mobile service solutions for student markets), but their contributions decrease in less familiar use scenarios (e.g., students generating ideas for mobile service solutions for business people). Contextual use experience is one of the capabilities that ordinary users contribute to NPD/NSD and which makes their contribution relevant.

Another contribution is their problem-solving approach. Ordinary users normally lack technical knowledge about a product's functioning and manufacturing processes. While this lack of technical knowledge has often been criticized as hampering ordinary users' ability to provide meaningful input for NPD/NSD (e.g., MacDonald, 1995; Veryzer, 1998), recent research alludes to this knowledge lack as a potential strength. Lack of experience with the technological side of product development stimulates users to approach innovation tasks more playfully and intuitively, thus triggering divergent thinking, which allows them to produce unique and novel ideas. Lay persons' more intuitive, unstructured, and flexible approach to problem solving liberates their divergent thinking capabilities. Intuition is "a cognitive process that somehow produces an answer, solution, or idea without the use of a conscious, logically defensible step-by-step process" (Hammond, 1996, p. 60). Intuition involves the tacit integration, balancing, and aggregation of various different informational cues. Research on dual processing theory regards intuition as an automatic, implicit system of thought in contrast to a more conscious, analytical, and explicit rational system (Kahneman, 1994; Lieberman, 2000). Generating RNP ideas requires exploration and a willingness to deviate from existing routines (Reid and de Brentani, 2004; Smith and Tushman, 2005). Intuitive information processing facilitates the execution of explorative activities (Leonard *et al.*, 2005).

In contrast, individuals with specific technological skills often use these skills in a convergent-thinking style more focused on the feasibility of ideas and their market potential (Magnusson *et al.*, 2010). Providing ordinary users with technical product information prior to an idea-generation task curbs their creativity (Magnusson, 2009). Moreover, a study comparing users with and without technical education (Kristensson *et al.*, 2004) substantiates the superiority of users without technical skills regarding generating highly novel ideas and customer benefits. In other words, lack of awareness of how current technology operates and fewer concerns about technical feasibility unleash ordinary users' divergent thinking, while convergent thinking about how an idea can be transformed into an offer hinders professionals' ability to generate completely new ideas.

Expert users have domain-specific skills that contribute differently to NPD/NSD than ordinary users' contextual use experience and convergent-thinking style. Expert users apply their profound knowledge of a specific area in which they have outstanding skills in order to solve tasks within this domain and to push its limits. If they have sound, domain-specific knowledge of an area that differs from a company's skill set, their integration helps the company access distant knowledge. While ordinary users' ability to contribute novel input is limited to providing rough ideas for future products, or new strategic arenas, expert users are in a better position to demonstrate their solutions' functional feasibility.

Research note: The different forms of expert users

Trait-based research highlights the relevance of domain-specific knowledge (i.e., the skills an individual has in respect of a specific domain) rather than overarching personality traits in the context of individual creativity (Csikszentmihalyi, 1994; Amabile, 1996; Taggar, 2002; Griffin *et al.*, 2009). With a focus on customer integration into NPD, the lead user is an intensively studied form of expert user.

Lead users are users who are ahead of a relevant trend and expect to benefit from an innovation that addresses this trend (von Hippel, 1986; Urban and von Hippel, 1988; Lilien *et al.*, 2002).

Extensions of the lead user construct comprise embedded lead users (Schweisfurth and Herstatt, 2015; Schweisfurth and Raasch, 2015) and technical lead users (Lettl *et al.*, 2006b; Lettl, 2007). Other expert users identified as contributing successfully to NPD are users with an emergent nature and technologically reflective users.

Embedded lead users are firm employees and simultaneously lead users of their firm's product and service offerings. Embedded lead userness is positively associated with gathering needs information from the market and facilitating its dissemination through internal boundary spanning behavior and domain-specific innovativeness (Schweisfurth and Herstatt, 2015; Schweisfurth and Raasch, 2015).

Technical lead users are users with lead user characteristics, who also have access to technological know-how and have technological competencies. This additional technological expertise means technical lead users can develop functional prototypes actively (Lettl *et al.*, 2006a, 2006b, 2008). For example, a team of neurosurgeons has developed a preliminary prototype of a computer-assisted navigation system for neurosurgery. One team member has a background in mechanics, having been a professional watchmaker before he started his career as a surgeon. Another surgeon is a professional, but auto-didactically trained, computer programmer. The

(*Continued*)

(*Continued*)

third one is an electronics expert (Lettl, 2007). Like technical lead users, users with high technical skills have been found to generate ideas with high technical feasibility (Magnusson, 2009; Wellner and Herstatt, 2014).

Users with an emergent nature are user types with distinct domain-specific skills, which allow them to imagine ways to develop concepts into promising innovations for the mainstream marketplace. These users have forward-looking and anticipatory competences and a high knowledge of product and service applications. This enables them to gauge a product or service's market potential from an ordinary user's perspective. They focus less on solving their specific problem than lead users do and show greater interest in understanding current market trends and developments, as well as in predicting future ones. These users have been found to be better positioned than lead users to develop RNPs that mass markets might perceive as valuable and wish to purchase (Hoffman *et al.*, 2010).

Technologically reflective users have a tendency to think about the societal impact of a technological innovation. They are strong in perspective taking (i.e., the ability to adopt other people's point of view (Davis, 1980)) and technical skills (i.e., the ability to solve technical problems (Benaroch *et al.*, 2006)). Such individuals are in a superior position to generate new product features and uses for technical products and to suggest technical solutions that help solve social problems (Schweitzer *et al.*, 2015).

Technically skilled users have the technical background required to provide feasible improvements to existing products that can be quickly integrated into existing product generations (Schweitzer *et al.*, 2014). While they are good at providing incrementally new product ideas, the opposite is true of more radical ideas. Technological expertise primes thinking for, or restrict solutions to, the current capabilities of the technological area to which their expertise applies (Frensch and Sternberg, 1989; Leonard-Barton, 1992) and inhibits divergent thinking, which is important for generating radically new ideas (Guilford, 1967; Amabile, 1996).

Watch this video for a great introduction into the lead user method: https://www.youtube.com/watch?v=tNKrX1QxN6U&list=PLD4C0E9AEDF085119 (accessed 30 August 2017).

This one provides further insights on technologically reflective users: https://www.youtube.com/watch?v=B1klMTNbLlo (accessed 30 August 2017).

Customers' motivation to be creative influences their creative outcome (Füller *et al.*, 2012; Mack and Landau, 2015). Close attention should be paid to the mechanisms that motivate participation. The quality and quantity of ideas gained through direct online approaches depend on the degree to which a company manages to attract a high number of submitters with a suitable background for the task (Füller, 2010; Amabile and Pillemer, 2012; Füller *et al.*, 2012). This holds true not only for online tools. The quality of direct offline approaches' output also hinges on customers' willingness to participate actively (Belk *et al.*, 2013; Tuckel and Wood, 2001).

Research note: Rewards, motivation, and the truth and myths of motivation crowding

Intrinsic and extrinsic motivation. Rewards, or incentives, can link to a range of motivations from more internal to extrinsic aspects. Participants are internally motivated to participate in an idea competition if they participate out of curiosity or the sheer enjoyment of participating. Competition designers have to pay attention to making participation experience playful, enjoyable, and inspiring to stimulate intrinsic motivation. In the context of great artistic achievements, performing creative tasks due to intrinsic motivation, such as the sheer interest and enjoyment, has often been associated with increased creativity. Artists with an interest in or an obsession with the creative act have been found more successful than those pursuing fame and fortune (Getzels, 1987; Collins and Amabile, 1999). Intrinsic motivation, is, however, never superior. For example, in the context of crowdsourcing, recent studies have indicated that in terms of successful contributions, external motivation is more relevant than internal motivation (Kosonen *et al.*, 2014; Mack and Landau, 2015; Landau *et al.*, 2016).

At the other end of the range, there is extrinsic motivation with direct compensation driving participation. This compensation can take the form of cash or rewards in kind, trophies, and awards. Internalized extrinsic motivations lie between the two extremes of mere intrinsic or mere extrinsic motivation. These motivations include altruistic motives, the wish to build and develop own skills and competences, the acquisition of valuable

(*Continued*)

(Continued)

information, and the search for skill development and self-efficacy. Awards and credits are incentives that address most of these motivations. These rewards also satisfy the participants' need by providing social motivations, such as recognition from peers, friends, or competition's organizers, as well as the motivations of self-marketers who want to demonstrate their skills and abilities. Exchange with experts is very valuable for participants striving to build their knowledge, (Füller, 2010).

As a tendency, extrinsically motivated individuals are more focused on the outcome of a crowdsourcing experience, while internally motivated ones consider the process of participating in the crowdsourcing experience more relevant (Hoyer *et al.*, 2010; Pera *et al.*, 2016).

Uses and gratifications approach. The uses and gratifications approach (Katz *et al.*, 1974) provide another way of structuring the benefits customers seek when participating in company-initiated creativity tasks. The framework focuses on nonfinancial benefits and differentiates between cognitive, social integrative, personal integrative, and hedonic benefits. Cognitive benefits are all the elements that help individuals learn, for example, improving their understanding of a product and its features. Social integrative benefits relate to the social ties that customers build through exchange with others and are manifested in a sense of social identity or a shared spirit. Personal integrative benefits are elements that help customers increase their status, credibility, and reputation, as well as provide a feeling of self-efficacy. Hedonic benefits are affective elements enabling customers to enjoy an experience (Katz *et al.*, 1974; Nambisan and Baron, 2009).

Rewards and types of participants. The choice of reward is closely related to the type of attracted participants. For example, ordinary users might mainly participate in idea competitions, or other manufacturer-initiated methods, to express their new product or service ideas due to their wish to see these ideas realized and to benefit from using the new offering. They may also enjoy participating as a way of relating to a brand with which they identify. Quite often, however, online idea competitions' participants are not ordinary users but those with domain-specific skills. Although they may possess domain-specific skills, they are not users or potential future users of the companies' offerings but instead fall into the category of

(Continued)

(*Continued*)

potential future employees or business partners. This means that the idea of using the product or service for which they provide information does not genuinely motivate them. Individuals with a professional background regarding a particular topic are typically interested in networking with an eye on future projects, or simply interested in turning their expertise into money. They are additionally often interested in the problem-solving process itself and have a genuine interest in the specific theme. Nevertheless, intrinsic motivation alone will not make them participants. They will pay close attention to IP issues and might even refrain from participating if they fear that they could accidentally produce confidential ideas that are not yet protected or if they find that the company behind the idea competition does not treat IP issues well. These individuals are interested in attractive cash prizes, in a percentage of the final offering's sales, or in awards that benefit their careers. Before choosing rewards, companies running idea competitions need to be clear whether they want to attract future customers in an endeavor to obtain insights into their needs, or as a means to engage with them, or whether they want to access their customers' creative potential. If the latter is the case, companies might have to provide background information on their customers and their needs in order to receive meaningful ideas, as the idea competition's target group is not equal to the final offering's target group.

Motivation crowding. The government of Switzerland has been considering Wolfenschiessen as a zone to deposit nuclear waste. Asked about their agreement with this plan, 50.8% of Wolfenschiessen's citizens agreed due to their national pride, public spirit, feeling of social responsibility, or expectations of employment. In a second survey, its citizens were offered a hypothetical CHF 5,000 in government compensation and their agreement with the plan dropped to 24.6%. As an external motivator, money reduced their internal motivation and was perceived as bribery. Similarly, participants in an experiment with a certain interest in solving puzzles received two different conditions. The experimental group was not paid on the first day but paid on the second day and then again not paid on the third day of the experiment. The control group was not paid on any of the 3 days. The experimental group was strongest motivated to solve the puzzle on the

(*Continued*)

(*Continued*)

second day, even working in breaks, but least motivated on the third day. This phenomenon is called an overjustification effect, which is similar to motivation crowding. When individuals are ready to act through internal motivation, monetary incentives suppress nonmonetary motivation. Once monetary incentives are stopped, individuals' prior intrinsic willingness to participate has also disappeared (Carlson and Heth, 2007). While this phenomenon is often used to call for smaller bonuses for sales managers, or as a reason for not giving children chocolate in return for "good" behavior, the overjustification effect only comes into effect if certain conditions apply: The intrinsic motivation is largely high, the rewards are independent of the receivers' competence, and the reward receivers perceive the reward as an authority's means of control. In other cases, extrinsic motivation could increase internal motivation (Rosenfield *et al.*, 1980; Eisenberger and Cameron, 1996; Cameron, 2001) and performance (Gerhart *et al.*, 2009).

The nature of a problem — its visibility, specificity, and complexity — influences creative customer actions. Some problems are easier to identify than others. The better users can identify the link between problems with a product and the impact of certain actions; the more they engage in cause and effect learning during their interaction with a product, the better they can apply this knowledge to creative problem solving. For example, when making pizza, the link between spreading the cheese evenly and the pizza's quality is more visible than that between preparing the dough and the pizza's quality. Identifying and solving problems is easier in the former. The latter requires more time and experimentation to get the product components and their interaction right, because the links between the action and the outcome are less tight (Lee and Walsh, 2016). Further, a problem can focus on a specific technical solution or on a search for more general suggestions for future activities. Ordinary users are generally better contributors to the latter, while expert users can provide insights into the former. Finally, product complexity decreases customers' ability to contribute.

Research note: Beyond creativity

Integrating customer directly in order to open the solution space not only generates creative input and novel solutions that fuel NPD/NSD but also has three distinct promotional effects.

The "I designed it myself" effect demonstrates that customers who are personally involved in cocreation activities identify stronger with the final offering and are willing to pay a premium for solutions that they cocreated. Self-design increases the subjective value and this effect is even stronger when customers design a product with a high subjective quality (e.g., if they like the design of that which they created). The larger the customers' contribution to the design, the stronger the link between cocreation and subjective value (Franke *et al.*, 2010).

The "customers @ the periphery" effect signifies that customers who do not participate in cocreation themselves, but learn that a company involves other customers, prefer this company's products to non-codesigned alternatives. In a series of experiments, Fuchs and Schreier (2011) showed the same T-shirt, bicycle, or piece of furniture to two different customer groups, once pretending the product was codesigned, once maintaining that corporate-internal designers designed it. Customers perceive cocreated products as demonstrating that such companies have a higher customer orientation, attach more favorable corporate attitudes to these companies, and show stronger purchase intentions. Customers prefer a codesigned product, because it makes them identify with the firm and feel more connected to it. In a follow-up, Dahl *et al.* (2015) identify customers' similarity to the cocreating customers as a boundary condition. For example, female customers prefer muesli cocreated with a 95% female community of cereal fans to muesli designed in-house, but prefer the latter to muesli cocreated with a 95% male community. The opposite holds true for male customers. In the same vein, customers at the periphery without software-developing expertise do not prefer cocreated video editing software to an in-house created software, while customers with a certain level of software-developing expertise strongly prefer cocreated software. Cocreation can also lead to negative reputational effects. For example, customers perceive luxury fashion items cocreated with customers as being of lower quality and as less prestigious than those designed without customer integration,

(*Continued*)

(*Continued*)

resulting in a preference for the latter. This research also indicates the differences between customer types. Fashion items are less preferred when ordinary users are integrated into codesigning them but are more preferred when customers with domain-specific skills (e.g., celebrities, artists, or other, specifically legitimized, users with recognized creative talent or creativity status) are integrated. The involvement of these customers with domain-specific skills seems to establish a certain distance from ordinary users, which is important enough for nonparticipating customers to believe that it affects luxury products' status and quality minimally (Fuchs *et al.*, 2013).

The "winning ambassadors" effect points to the active role that customers integrated into the NPD/NSD consciously or unconsciously play in promoting these offerings. Case study evidence indicates that engineers, who had served on the advisory board of a radical NPD/NSD project and had worked in a potential customer company, later encouraged their companies' buying centers to adopt the innovation (Abetti, 2000).

Integrating customers has psychological effects not only on customers but also on the manufacturers' employees. A **feeling of reciprocity** is a positive effect. In an experiment, cooks made tastier food when they saw their customers, and customers rated quality higher and were more patient when waiting for a meal if they could see the cooks. The relevant study rules out stress or accountability as explanations for this behavior, finding that reciprocity and the feeling of being appreciated are the central drivers of the cooks' ambitions to create tastier dishes when they can see their customers (Buell *et al.*, 2016).

A negative effect is the **not-invented-here (NIH)** syndrome, an irrational devaluation of external knowledge merely because it is based on external, although it might be valuable. Such negatively shaped discriminatory attitudes are often rooted in the fear of losing influence, status, or rank due to the external impetus. For example, R&D managers who are asked to generate ideas themselves might fear that their seniors think they do a bad job if they did not produce an equally good idea and might thus turn their back on the externally generated idea (Katz and Allen, 1988; Antons and Piller, 2015).

For more on NIH and ways to overcome it, watch https://www.youtube.com/watch?v=LqGWGUaWJ3A (accessed 30 August 2017).

2.2 Direct online tools for opening the solution space

Over the last two decades, ICT development and customers' increasing online connectedness have led to a surge of companies integrating customers into NPD/NSD by means of direct online tools (Füller, 2010; Mühlbacher *et al.*, 2011; Schenk and Guittard, 2011; Bayus, 2013). Direct online tools enable a broad search for new ideas by giving their users access to high numbers of idea providers and far-away customers with diverse backgrounds (Bullinger *et al.*, 2010; Boudreau *et al.*, 2011; Jeppesen and Lakhani, 2013). Conversely, non-ICT-based tools entail more local searches with more familiar idea providers (Füller and Matzler, 2007; Droge *et al.*, 2010; Schweitzer *et al.*, 2012; Nishikawa *et al.*, 2013; Natalicchio *et al.*, 2014).

Direct online tools allow the inclusion of groups around the globe, thus moving beyond local searches and attracting problem solvers from the periphery (Prandelli *et al.*, 2006; Obrist, 2007; Bilgram *et al.*, 2008). These tools also facilitate knowledge transfer (Hemetsberger and Godula, 2007), and from the idea submitter's perspective, participation in online competitions is relatively easy, as the tools allow access at any time and from anywhere (Schweitzer *et al.*, 2012; Gassmann, 2013). Consequently, direct online tools facilitate the gathering of fresh ideas. In addition, they have the potential to decrease the time and effort needed for customer integration. Innovation initiatives, which used to take months to coordinate and launch with non-ICT tools, such as cocreation workshops, can be started at comparably short notice.

Direct online tools comprise central crowdsourcing tools. Crowdsourcing is formed by the words "crowd" and "outsourcing," thus describing the broadcasting of a business task to the mass of people accessing the World Wide Web to find if any of them can solve this task. Crowdsourcing builds on a self-selection principle according to which the most apt individuals for solving the task will participate, because they have the intrinsic motivation to tackle the task, and their problem-solving costs will also be lower, because they have the appropriate knowledge and means to solve the problem (Geiger *et al.*, 2011; Kosonen *et al.*, 2014).

Research note: Transaction costs, local search bias, information stickiness, and direct online tools

Transaction cost theory investigates ways to minimize transaction costs. Transaction costs are all the costs associated with transferring property rights between entities. These costs include preparing a transaction (e.g., the research and potential partner evaluation costs), carrying out a transaction (e.g., negotiation and process coordination costs), and control (e.g., the quality controlling and timing costs). Transaction costs depend on the type of transaction and the way it is organized (e.g., through the market, partner networks, or internally in the company). The higher the task specificity and the task uncertainty (i.e., the likelihood of adaptations and changes throughout the task fulfilments), the better it is to perform a task internally. The reasons for this is that the costs of assuring the quality of these tasks, of exchanging highly specific knowledge, and of adapting dynamically when the requirements change throughout the process, are lower in this organizational form (Coase, 1937; Williamson, 1985). ICT technology changes the different organizational forms' transaction costs by decreasing those of market solutions and of hybrid solutions for higher degrees of task specificity. This decrease is indicated by the move from X_1 to X_1' and from X_2 to X_2' in Figure 13. In other words, ICT facilitates the exchange of information with high task specificity and task uncertainty. ICT reduces

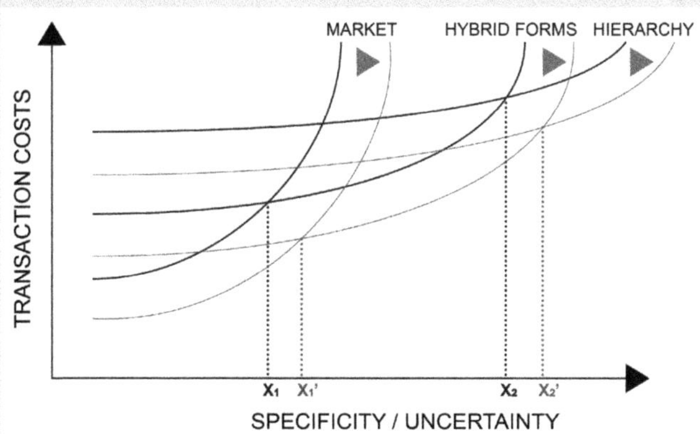

Figure 13: The influence of ICT on transaction costs in different organizational forms. (Adapted from Picot *et al.* (2003) and Reichwald *et al.* (2009).)

(*Continued*)

(*Continued*)

the transaction costs of all organizational forms but proportionally more of market transactions and network transactions, because ICT facilitates and accelerates the search for relevant market offers, and the exchange of information between geographically distant contractual partners is easier with than without ICT (Malone *et al.*, 1987; Reichwald *et al.*, 2009).

By reducing the costs of distant search, ICT enables the accumulation of distant knowledge. Direct online tools allow increasingly distant customers to provide input for NPD/NSD, thus increasing the innovating company's likelihood of receiving novel insights. Put differently, online tools reduce the local search bias, which describes individuals' tendency to draw on past experience and locally available information to solve tasks. Individuals tend to apply well-known and proven methods, as well as routines that fit the existing spectrum of knowledge, thus applying only a limited solution space within the near area of existing knowledge (Katila and Ahuja, 2002). Local knowledge search magnifies the chances of developing workable solutions but reduces opportunities for finding radically new solutions (Rosenkopf and Nerkar, 2001; Fleming and Sorenson, 2004). Distant search provides access to knowledge residing beyond this local area of knowledge, thus opening the solution space to novel knowledge recombination and to discontinuous innovations. Direct online tools reduce the transaction costs of accessing distant knowledge by offering relatively fast and cheap access to a large amount of potential solvers with distant knowledge (Afuah and Tucci, 2012; Piezunka and Dahlander, 2015).

Furthermore, ICT technology has the ability to reduce information stickiness. Information stickiness is the "incremental expenditure required to transfer that unit of information to a specified locus in a form usable by a given information seeker" (von Hippel, 1994, p. 430). When information is sticky, the transaction cost is high. Information stickiness is an equation consisting of the type of information (e.g., the more implicit and highly specific the knowledge is, the greater the stickiness) and the receiver of the information (e.g., the lower the receiver's relevant qualification and previous knowledge, the higher the stickiness). ICT reduces information stickiness when it allows transaction partners to transform implicit, tacit knowledge into tangible artifacts. For example, instead of asking customers to explain what T-shirt designs they like, the innovating company can

(*Continued*)

(Continued)

provide online tools that allow them to design their T-shirt themselves, thus reducing iterative slopes of design optimization between the manufacturer and the customer. Prerequisites for this stickiness reduction through ICT are (1) that the tools required to empower customers to make their implicit needs explicit should be available, (2) that the customers have the necessary skills to use these tools accurately, and (3) that the value-creating task of producing the final solution can be partitioned meaningfully between the customers and the manufacturer. Task partitioning into clearly definable and exchangeable subtasks and the visibility of the tangible contribution might be easy for T-shirt designs, or for software codes, but more challenging for highly interdependent technological solutions, such as the development of medical equipment.

These advantages induce many practitioners to try direct online tools. Many initiatives, however, underperform or fail (Dahlander and Piezunka, 2014). Proactive attention, such as submitting a company's own suggestions via the online platform, reduces failure, because the suggestions give the idea contributors information about the type of ideas that are welcome, decreases the risk of the contributors misunderstanding their expected contribution, and helps build trust. Further proactive attention allows external contributors to see how the idea is further processed, provides feedback on an idea, and provides improvement suggestions. This mutual interaction gives the idea contributors a reciprocal partnership role. Conversely, a one-sided submission of ideas without interactions makes contributors perceive their role as that of a cheap source of new ideas, thus reducing their motivation to contribute (Harhoff *et al.*, 2003; Shah, 2006; Dahlander and Piezunka, 2014).

Proactive attention is especially important in the early stages of online idea platforms until initiatives have sufficient momentum to become self-sustaining. Paying attention to contributions is another important way of increasing online platform success. Participants need to feel that their input matters, because if this is scorned or ignored, it may lead to bad word of mouth about the initiating company. Idea contributors want the initiating company to acknowledge their idea, to invest effort in considering its value and to keep them informed about the current status and about whether it will be realized or not. Clear and transparent communication on these aspects is very important for idea contributors who submit an idea to a specific platform for the first time (Dahlander and Piezunka, 2014).

The potential downsides of customer integration via direct tools are that customers might be reluctant to participate, might hijack the company's initiative for their own ends (e.g., by using it as a vehicle for criticizing the company, or by providing, or voting for, ridiculous ideas to make fun of it), or might provide mediocre ideas that do not reflect substantial input for truly new and radical innovations, because these ideas lack originality, feasibility, and value (Shah, 2006; Piller and Walcher, 2006; Knudsen, 2007; Rushkoff, 2007; Pisano and Verganti, 2008). A customer's use experience and personal preferences may result in new product or service ideas that are too specific, or uninteresting, for the company's wider range of consumers (Antorini and Muñiz, 2013). Further, the costs, time, and resources required to screen the large, ICT-triggered influx of customer-generated ideas might reduce the available resources for other innovation endeavors (Gupta and Wilemon, 1990; Nambisan, 2002; Bonner, 2005; Allocca and Kessler, 2006; Laursen and Salter, 2006; Toubia, 2006; Harmancioglu *et al.*, 2007; Dahan *et al.*, 2011; Rubera and Kirca, 2012; Soukhoroukova *et al.*, 2012). R&D employees who are usually responsible for generating ideas could also oppose externally generated ideas (NIH syndrome) or hinder the company's ability to act upon such ideas, because these ideas are too peripheral to fit these employees' current solution base (lack of absorptive capacity). These internal hurdles and potential remedies are discussed in detail in Chapter 6.

Generally, direct online tools gather user ideas online. In some cases, users can also read other users' ideas. Either only registered users can access these ideas to comment and build on them or an unidentified mass of users can do so. While such access can cross-fertilize idea generation, it comes at the risk that competitors may also access and learn from the proposed ideas (Ebner *et al.*, 2009), thus adding to the general danger of information leakage and intellectual property quarrels, which might arise when customers are integrated as solution providers (Brown and Duguid, 2001; Enkel *et al.*, 2005; Bartl *et al.*, 2012). Furthermore, submitted ideas might already be under development in the company. Once the company launches such a product, the customer who submitted the idea might consider the idea stolen and claim property rights. Clear corporate policies can reduce legal problems; nevertheless, the problem remains that such actions could lead to disappointed customers turning their backs on the company. More generally speaking, a company reveals sensitive information about its innovation strategy and activities by broadcasting an idea competition with a question or a problem to solve. Competitors are aware of this and could use it for their own ends. Moreover, an idea that has been publicly published runs into difficulties when a patent is sought. Therefore, the initiating company should scrutinize the scope of the direct online tools it applies. For example, it must decide whether everyone should be able to see submitted ideas and comment on them or whether the ideas are only transferred by means of an online toolkit but are not visible online (Alexy *et al.*, 2011; Gassmann and Schweitzer, 2014).

2.2.1 *Goals and scope of direct online tools*

Online direct tools need appropriate planning and execution of their scope to deliver the desired benefits and avoid drawbacks. The scope of direct online tools depends on the goal of the customer integration activity (Table 2).

Table 2: Strengths and challenges of direct online tools.

Strengths	Challenges
• Access to distant knowledge	• IP issues
• Transfer of sticky information	• Spying by competition
• Relatively cheap and fast way of acquiring creative input	• Difficulties of absorbing new ideas
• Inspiration of contributors through other contributors' submissions	• Risk of NIH syndrome
• Possibility to not only gather ideas but also receive comments and other feedback on ideas	• Risk of deliberate or unintended submission of useless contributions

The goal describes the expected degree of the innovation project's innovativeness (ranging from incremental to radical in terms of the technology and market) for which the customer integration project needs to provide ideas and the type of information sought (insights into customer needs (= looking for a customer problem to be solved) versus creative solutions for a technically challenging problem (= looking for a solution to a problem)).[1] A clear goal description is critical, because it calls for a different arrangement of the elements that build direct online tools' scope. This scope includes all the elements that need to be specified before idea gathering via direct online tools can start. It comprises the degrees of idea elaboration and task specificity, the choice of tool, the toolkit and its online presence, the type of participants, their empowerment, as well as the way of reaching, communicating with, and motivating the participants (Figure 14).

The *degree of elaboration* describes the required thoroughness and detail with which the company expects a user to submit an idea. A clear definition of the degree of elaboration is important for the user and for the idea evaluating committee. A low degree of elaboration (e.g., an idea described in a few words or sentences) attracts more participants than a high degree of elaboration (e.g., detailed functional sketch or a functional prototype), as the latter requires domain-specific skills and a higher investment in time and effort.

Regarding the *choice of tool*, the options are online idea competitions and virtual idea communities. Online idea competitions are events with a competitive character that run for a predetermined period to gather ideas on a particular topic. Virtual idea communities are ongoing platforms on which users can post ideas.

The *choice of toolkit* ranges from toolkits that only allow the transfer of solutions in a constricted and narrow format (i.e., customer can choose from a set of different colors and flowers to codesign a label) to toolkits

[1]We focus on innovation management goals. The communication goals of applying the above online tools would refer to customer relationship building, image building, or promoting a certain brand. We touch on this issue in the box *Research note — Beyond creativity* in Section 2.1.

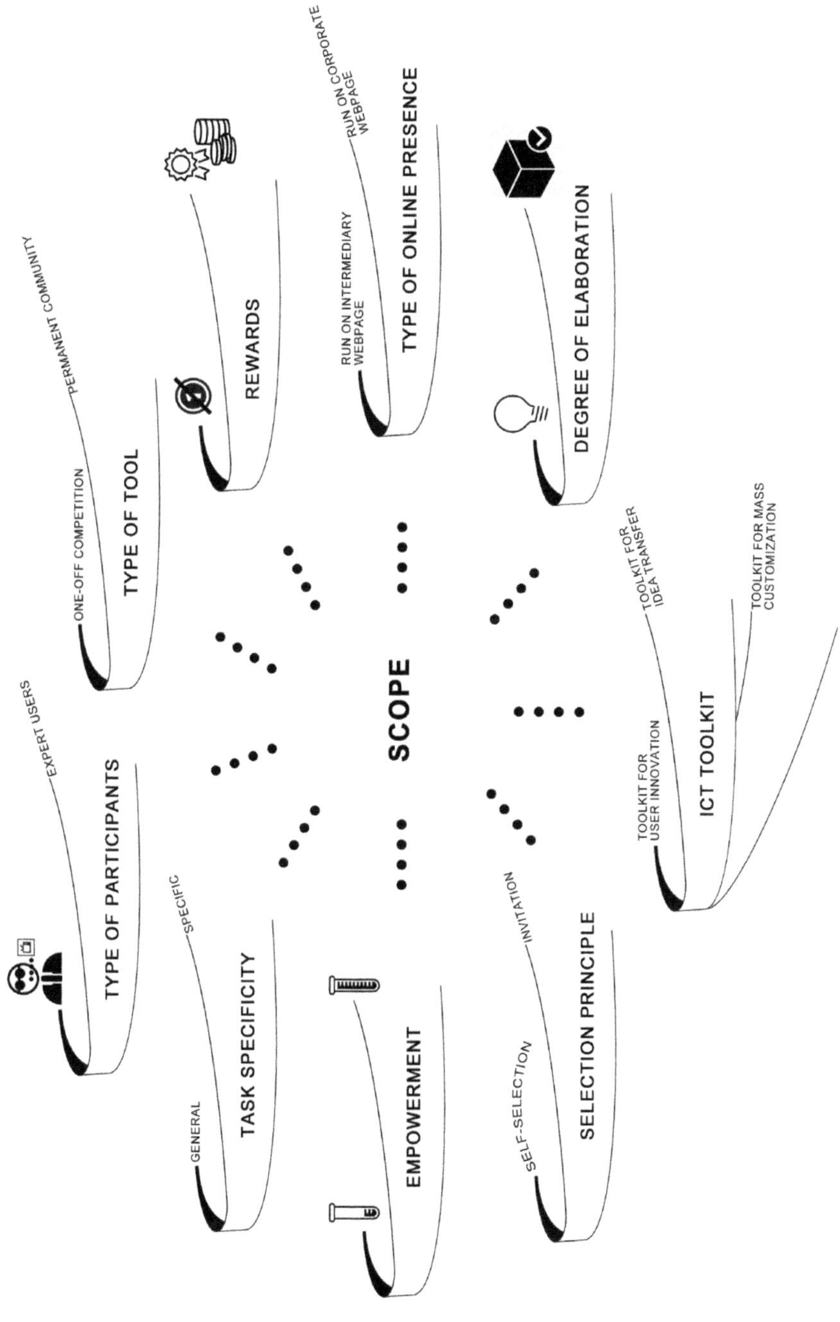

Figure 14: Scope of direct online tools.

with more open solution spaces (see the box *Customer integration in action*: *Toolkits*, *mass customization*, *and OI*).

Customer integration in action: Toolkits, mass customization, and OI

Toolkits are tools that a company provides to encourage idea transfers between the customer and the company. Toolkits enable customers to create a new idea, propose a solution, or configure their ideal product or service. Toolkits take the following three forms (von Hippel and Katz, 2002; Piller and Walcher, 2006; Reichwald *et al*., 2009):

1. Toolkits for mass-customization application allow customers to generate individualized products for their own use. These toolkits support value-added activities in the production process by allowing customers to choose their favorite features from a predefined set of features. The solution space is restricted and the customer's individual design is not used to feed the innovation funnel but to produce a customized product for this user. Mass-customized products often sell at premiums of 30–50% on comparable standard items. In the fashion industry, such toolkits allow customers to, for instance, select their favorite color combinations. In the B2B world, a famous example is IFF, a manufacturer of fragrances and flavors that allows its customers to experiment with its substances in order to create various scents and tastes, after which they can order their favorite combination.
2. Toolkits for user innovation aim at value-added activities in the innovation process. Customers create products with such toolkits and submit them to contribute to NPD by potentially creating a product that suits the taste and interest of a wider group of customers. The solution space is wider but still limited to a specific range of features or designs that customers can create. Lego's digital toolkit is such a toolkit that allows its customers to design any shape of brick creature as long as it is based on existing Lego bricks. Before 2011, it included a "Designed by Me service," which allowed the designer and other interested customers to buy the designed product. In 2011, Lego stopped this service due to its high complexity and its failure to meet quality expectations. Customers can still design new sets with these toolkits and submit them to the Lego Ideas platform. Internal committees later scrutinize ideas that receive a certain number of votes from the platform members.

(*Continued*)

(Continued)

3. Toolkits for idea transfer provide an unlimited solution space that helps customers convey and transfer any idea. For example, the InnoCentive platform's idea submitters are allowed to design ideas or concepts with any software program, develop physical prototypes, send pictures of these, or provide handwritten and scanned construction plans, and then upload them together with their specifics as prescribed by the competition to which they are submitting them.

The *choice of an online presence* is whether this presence is on a webpage hosted by an intermediary, or on the idea-seeking company's webpage. Company-run initiatives usually attract submitters from the company's close environment, notably from its stakeholders, such as customers, suppliers, or companies and individuals interested in some type of exchange with the company. Intermediaries usually have a lively solver community not connected to the company that runs the initiative. For example, InnoCentive is a platform that allows others to publish their problems. In 2015, it had an active solver community of 365,000 registered solvers, more than 2,000 posted challenges, and more than 40,000 solution submissions. The financial rewards offered for these submissions are often considerable, ranging from $500 and $1 million, depending on the problem complexity (Innocentive, 2015). The average time needed to solve a problem is a staggering 74 h with 70% of the solutions originating from another scientific discipline or technical field than the field related to the posted problem. The speed at which problems are solved and the solutions' distance from the discipline indicate that even before the initiator posted the challenge, these solutions were already known as a by-product of other research but without knowing how this could be transformed into something useful (Lakhani, 2009). Moreover, socially marginal individuals — in the sense that they are distant from their professional community — have turned out to be the best solvers of such challenges. Intermediaries are thus a good way of accessing distant knowledge from experts with relevant domain-specific skills (Jeppesen and Lakhani, 2013).

Task specificity defines the topic of the idea generation task. Task specificity ranges from finding a solution to a precisely described specific problem (e.g., how can we make this smartphone lighter?) to finding general ideas for improving a product or service (e.g., in the form of potential problems/needs) that warrant future innovative activities (e.g., what should smartphones look like and be able to do in 1/5/10 y?). The more general the task, the greater the submissions' variety.

The *type of participants* ranges from only current customers to potential future customers to users with domain-specific skills. If the customer integration's goal lies in the problem identification field, a viable option is to draw on potential future customers by means of idea competitions with low task specificity. If a company focuses on improving its current offerings, the most appropriate approach might be to involve its current customers through virtual communities on which they can post problems with the current offerings and suggestions for improving them. If a company strives to find solutions that include highly innovative features, they should consider involving users with domain-specific skills. The input's innovativeness will be lower when familiar current customers are integrated and higher when distant users with diverse backgrounds are. Distance, however, causes absorption and integration problems; therefore, integrating very distant users can impact the ability to utilize the gained novel input (Cohen and Levinthal, 1990; Lee *et al.*, 2015).

As a *selection principle*, or way to reach the right type of participants, companies can either build on the right people's self-selection through the choice of rewards they offer, through their choice of the ways they communicate the idea selection tool, and through their choice of an online presence. In addition, the company can define criteria that the participants have to fulfill to qualify and then search for and invite such individuals. The disadvantage of the latter approach is that it might lead to local search bias, while the disadvantage of the former is that the "wrong" crowd might be attracted.

As far as *empowerment* is concerned, companies can involve customers online by asking them to submit ideas (create empowerment) and/or vote for ideas (select empowerment). Submitting ideas is a direct approach to open the solution space, while voting is basically a direct approach to closing the solution space. Companies often design online tools for selecting and creating empowerment. If customers are given full create and select empowerment, companies run the risk that they might hijack an online tool for their own purposes. Full empowerment signifies that only the customers create ideas and only the customers decide which ideas to realize. Precreate and preselect empowerment gives companies more decision freedom. Customers thus suggest designs and vote for designs, the companies in turn consider the designs and the choices, but the final decision regarding the products remains with the companies. Companies have to be very clear from the start about the degree to which they wish to empower their customers; changing empowerment halfway through a process can lead to frustrated customers (von Hippel, 2005; Fuchs and Schreier, 2011).

Through our interactions with industrial practitioners, we have gained the insight that the choice and combination of elements are what make the scope of direct online tools successful. Figure 15 depicts two paths that have been proven highly effective. The first path gathers ideas that represent insights into customer needs. To gather inspiration for potential future solutions to customer needs, direct online tools with low task specificity

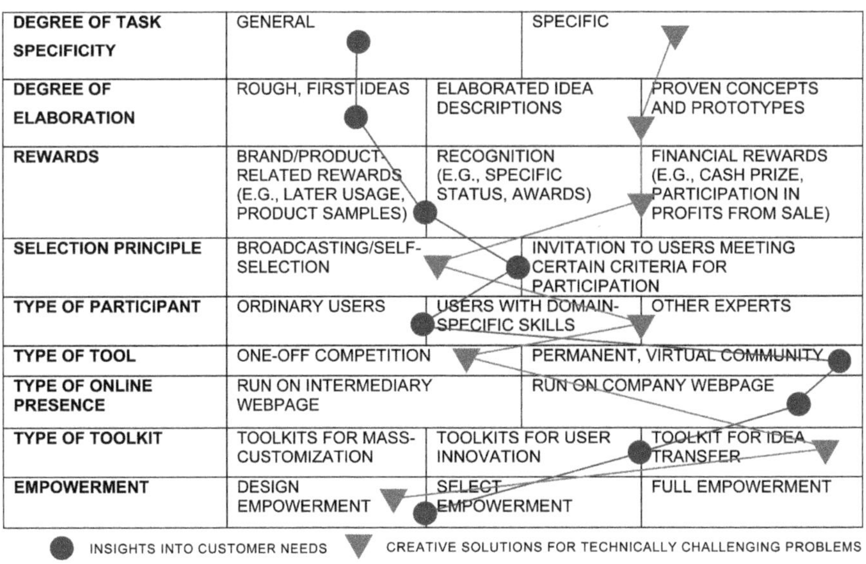

Figure 15: Successful paths of direct online tools along the scope.

and a low degree of elaboration allow ordinary users to provide ideas that uncover their current needs and problems. Continuous interaction with these users through virtual communities enables customers to receive permanent feedback on current product features and ideas for improvement. The second path is more suitable for technically challenging problems. In this case, companies strive to access distant knowledge from persons with domain-relevant expertise through one-off competitions focusing on a specific task and actionable concepts or prototypes.

2.2.2 *Online idea competitions*

In idea competitions, a company asks customers to submit ideas on a specific topic. Idea competitions run for a predetermined period, usually 4–30 weeks, after which a board usually evaluates the ideas and selects the winners. Online idea competitions are one-time, single, or multistage, events for gathering fresh ideas on specific topics (Piller and Walcher, 2006; Terwiesch and Xu, 2008; Athaide and Klink, 2009). Such idea competitions can be repeated at regular (e.g., annual intervals) or irregular intervals.

Customer integration into action: One-time contest and multistage tournaments

Cisco's first I-Prize competition took place in 2007. Building mainly on word of mouth, the competition received 2,500 idea entries from users, start-ups, students, and other external participants within 5 weeks. In a second step, Cisco invited 450 selected providers of idea entries to a virtual pitch. In a third step, 12 of providers had the opportunity to present their ideas to the top managers, who selected the final winner. The winner received a USD 250,000 cash prize and was offered position in a new business unit dedicated to realizing the chosen idea (Schweitzer *et al.*, 2012). Cisco repeated the I-Prize competition three times over the next few years and then changed it into the Cisco Grand Challenge (see Figure 16) with a focus on ventures as idea providers. The change indicates that, from the company's perspective, submitters with domain-specific knowledge rather than ordinary users are

(*Continued*)

(Continued)

Figure 16: Cisco's innovation grand challenge.

Source: http://www.innovationessence.com/innovation-grand-challenge/ (accessed 30 August 2017).

the most attractive participants, as they open the company's perspective to distant search fields.

The Starbucks-sponsored Betacup challenge was a successful one-time contest to gather ideas for reducing nonrecyclable cups by increasing the attractiveness of alternatives. The challenge ran in 2009 and received 430 entries, with the Karma Cup entry (see Figure 17) gaining the first prize (USD 10,000). The winner was not an idea for an actual cup but for a chalkboard registering every tenth person ordering a drink in a reusable cup, who would then receive a free drink (Evans, 2010). The challenge shows that online contests have the potential to provide innovative ideas beyond the company's immediate search frame. Starbucks had looked for a cup but found a service innovation. Starbucks partially realized the ideas by developing a recyclable cup and providing every customer bringing this cup along, or buying it at the shop and ordering coffee with it, an immediate cash reduction on the coffee price. This solution was introduced in specific country markets and is still currently available.

Idea competitions, however, sometimes only generate sparse activity. For example, SAP's idea competition SAPiens focused on selected groups

(Continued)

(*Continued*)

Figure 17: The winner of the Betacup contest.

Source: https://www.fastcompany.com/1661306/winner-starbucks-coffee-cup-challenge-isnt-coffee-cup (accessed 30 August 2017).

of German students, with 220 registered users submitting 70 ideas. The competition used an existing community and tried to stimulate more activity in this community through the competition (Ebner *et al.*, 2009). A better approach to promote a competition might be to build on a set of lively communities with the help of external intermediaries.

Online idea competitions comprise three phases. First, they have a preparatory phase that clearly defines the competition's goals. These goals influence the central elements of the idea competition's design, including the definition of the type of customers to integrate, the degree of elaboration, and the task specificity. A company aiming for a feasible solution to a technically challenging problem will specify the task to a high degree (e.g., coatings with superior abilities to direct liquids across flat surfaces to be used as containers' internal coatings in order for their content to flow out easily and completely) and require a high degree of elaboration in turn (e.g., a functional prototype). In order to receive useable submissions to meet this aim, companies must strive to attract expert users (e.g., users with a background in chemistry, or material sciences, who have a small laboratory) through an appropriate choice of communication channels (e.g., conferences and special interest media targeted at material science specialists, a clear statement on their webpage that they are looking for expert contributions), and rewards (e.g., financially attractive awards). Idea providers to online idea competitions are not necessarily just potential future customers but could include other external individuals, such as experts from other industries, or potential future employees. Companies looking for immediate solutions for tricky technical challenges are usually more successful in their external search when they move beyond their customer focus by including potential future employees and professionals with domain-relevant skills (Lakhani and Jeppesen, 2007; Natalicchio *et al.*, 2014).

On the other hand, online idea competitions restricted to customers have the potential to provide insights into latent needs and to serve as a basis to transfer sticky needs information (di Gangi *et al.*, 2010; Füller *et al.*, 2011; Schweitzer *et al.*, 2012). However, these ideas usually require further development, refinement, and translation in order to transform them into functioning and useful offerings (Nishikawa *et al.*, 2013). Companies looking for inspirational ideas and latent needs information from ordinary users will choose different ways of communicating the idea competition than when they target expert users. For example, in the former case, it is more appropriate to advertise the competition via mass media, Facebook fan pages of the brand, and customized communication to selected users representing an important customer segment.

In the implementation phase, the project is live. The company needs to attract idea submitters and provide them with information on the topic and the competition's goal, the expected task's content and degree of elaboration, the legal terms and conditions, the criteria that will be used to evaluate the ideas, the reward mechanism in place for the submitted ideas, and the number and composition of the evaluation committee members (Füller *et al.*, 2008; Landau *et al.*, 2016). In the implementation phase, the ideas are simply transferred from the submitters to the company; alternatively, the ideas can be published on the company webpage where registered customers, or an unregistered crowd, can study them. The webpage can display ideas to inspire potential participants to provide their own submissions, to comment on and refine others' submissions, and to vote for their preferred submissions. The upside of publishing submitted ideas online is that publication allows potential submitters to build on other submitters' ideas, to refine them, and use them as inspiration for new ideas, while the downsides include problems with IP rights and the risk of competitors studying and using these ideas.

In the evaluation and execution phase, a committee evaluates the ideas and selects those that receive a reward. The selection can include the participants' votes as an evaluation criterion. The evaluation criteria have to be clear from the start of the competition. To guarantee fairness, the evaluators should use a transparent process that respects the evaluation criteria. Participants have to be informed about the results in a timely manner as well as about further decisions and steps the company will take in connection with the submitted ideas. The evaluators should be granted sufficient time for their evaluation. A high workload hampers the evaluators' capabilities to address each idea adequately and encourages routinized and simplified decision-making. As a consequence, a high workload encourages the selection of incrementally new ideas that fit established repertoires and discourages the selection of radically new ideas (Csikszentmihalyi, 1996; Sullivan, 2010; Dahlander and Piezunka, 2014; Piezunka and Dahlander, 2015).

2.2.3 *Virtual communities for idea suggestion and cocreation*

Virtual idea communities are the basis for permanent online interaction with the company's customers, because they run continuously to allow customers to provide ideas, inform companies about difficulties or problems with an offering, or comment on others' input. Customers provide ideas repeatedly in online communities. Prior research indicates that the value of customers' input in the form of ideas posted in virtual communities varies over time. This means that they generate more repetitive and less diverse ideas over time. However, this effect is less strong with regard to idea submitters who comment on a very diverse set of ideas, or on other customers' suggestions (Bayus, 2013). In other words, customers seem to be cognitively fixed (Duncker, 1945; Smith *et al.*, 1993; Burroughs *et al.*, 2008), but context-shifting through diverse interactions with others can reduce the potential negative effect of cognitive fixedness (Smith, 2003).

Virtual communities have the potential to create customer–customer and manufacturer–customer interactions and relationships that are beneficial for NPD/NSD. In virtual idea communities, customers are typically free to pose any idea or suggestion that they consider worth communicating to the company or other customers. This interaction needs professional technical realization (e.g., easy access to the community via different devices, such as mobile phones, iPads, and laptops) and professional management. A virtual community manager has to keep the community growing and keep its participants active by engaging with the community members, providing feedback on submissions, creating gratifying and enjoyable experiences, asking for input on specific topics, and supervising discussions to avoid inappropriate content (Füller *et al.*, 2007; Porter *et al.*, 2011; Dahlander and Piezunka, 2014).

Community managers can keep a community active by initiating discussions or inviting members to lead such discussions. Individual employees in the My Starbucks Idea community successfully engage with community members by directly addressing them, thanking them personally for submitted ideas, explaining which ideas they implemented successfully, and asking for support for a specific planned activity. Such personal interactions create a feeling of affiliation and personal attachment. Employees and customers in this community also actively use this opportunity to create personal online profiles that provide room for self-expression and background information on their lifestyle and socio-demographics.

Many companies also give their community members different types of membership status, with the most loyal and active members usually receiving exclusive rights and other privileges. For example, Chick-fil-A, an American fast food chain, invites loyal and active members to join their Champions Panel. These members have the opportunity to try new menus before their market introduction and provide feedback. Such activities bind special status members even more to the company, give them a sense of status and the feeling that they can influence the company's actions, and the management often has more confidence in the quality and relevance of these members than in those of the average members (Porter *et al.*, 2011).

Research note: The value of serial ideators

Dell, a leading US PC maker and provider of tech solutions, launched the platform IdeaStorm (see Figure 18) at the beginning of 2007. The aim was to integrate its customers into the innovation process by providing its virtual community members with an opportunity to post, vote for, and comment on ideas. Over the last 10 y, its customers have submitted more than 25,522 ideas of which Dell has realized roughly 550.

Some research indicates that the low number of realized ideas (roughly 2.2%) may be due to Dell's lack of interaction and feedback on submitted ideas in the community's early years. This lack discouraged eager contributors, who turned their backs on the organization (Porter *et al.*, 2011). An alternative explanation can be found in the work by Bayus (2013), who maintains that ideators' functional fixedness is a reason for the rather low number of ideas that is worth realizing. Ideators who propose several

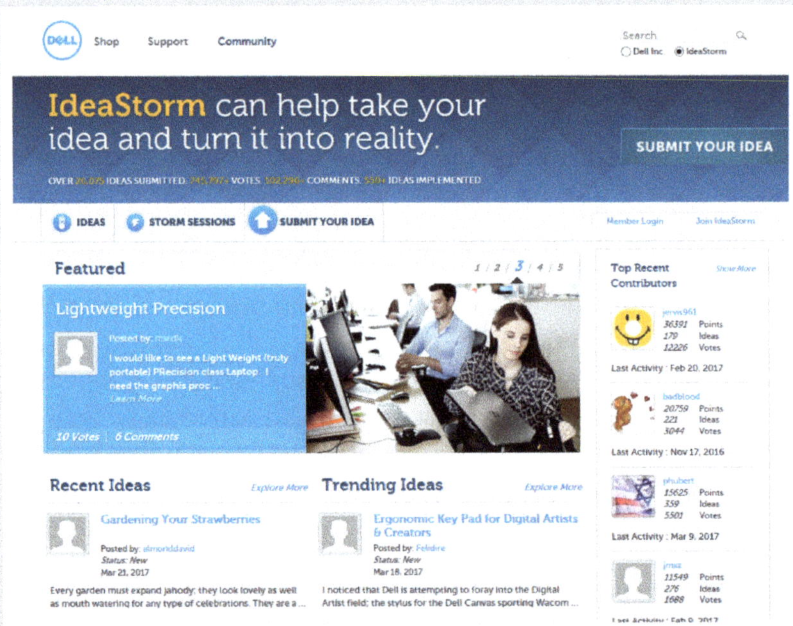

Figure 18: Screenshot of Dell's IdeaStorm community platform.

Source: http://www.ideastorm.com/ (accessed 30 August 2017).

(Continued)

(Continued)

ideas over time are more likely to produce an idea that Dell realizes than those community members who only produce one idea. However, ideators with one realized idea are less likely to produce another hit, because they tend to propose ideas in the proximity of their realized idea. Instead of producing a high quantity of diverse ideas, they are functionally fixed on a specific type of idea. This lack of repeated success is smaller in respect of ideators who not only propose ideas but also comment on other members' ideas. The reason for this phenomenon is probably because such ideators are in a better position to reduce their functional fixedness due to their openness to other members' views, opinions, and ideas. Dell has partially implemented processes to counteract functional fixedness problems. These processes include encouraging comments, running short-term idea-generation sessions on a particular topic of relevance (Dell Storm Sessions), as well as posting ideas already featured on their webpage and perpetually altering these to entice returning ideators to think about new topics.

While the majority of virtual communities for NPD focus on ideation, some move beyond mere idea generation and commenting platforms to create more holistic cocreation platforms. For example, the Lego Ideas webpage not only offers users the opportunity to upload ideas for a specific new brick model, but by building the model virtually, it also assures users that the building plan and the required bricks are transparent and easily accessible if the company makes a positive decision regarding manufacturing the model. Similarly, Quirky's intention regarding user interaction had clearly been cocreation activities, which included not only ideation but also concept and product development, as well as the associated marketing activities (see *Customer integration in action*: *The rise and fall of Quirky* for details). These examples of holistic platforms provide evidence of ambitions to entice customers to engage in concept development activities. Customer integration through virtual communities can also contribute to NPD/NSD in the later stages of the development process. For instance, many B2B companies and software companies provide support-focused virtual communities in which users can exchange their product problems. These communities can be interpreted as codevelopment platforms that enhance the product, its utilization, and its value through a product-service offer. For example, MakerBot, a manufacturer of 3D printers, hosts a vibrant online community that exchanges digital image files allowing users to search for designs they like and to print them, which increases the value of having a 3D printer. In addition, users who generate designs can either offer designs for free, or sell them, which partially turns these users into cosellers (Gilleran and Noyes, 2015). Such user-generated product enhancements can also occur unintentionally. In 2010, a user-generated webpage with illegal hacks by Kinect, a device for the Xbox that controls video games via body movements, went live a few days after the device's launch. Hackers have developed drivers that allow a camera and depth sensitivity functions to open Kinect's 3D motion-scanning features for applications beyond gaming. Kinect has been used for virtual-reality-enhanced fitting rooms, as a gestural interface for medical instruments, as art projects, and many other applications. While Microsoft's first reaction was to sue these users, the company later supported Kinecthak.com and other sites that share hacks that enhance Kinect's utility and actively search for Kinect's application in new business fields. Kinect eventually made it to the Guinness book of record as the fastest selling electronic consumer device with eight million units sold in the first 60 days after its launch (Johansen and Ronn, 2014; Wikipedia, 2017).

Research note: Open source and open design communities

Users not only exchange ideas or cocreate solutions in company-initiated communities but also innovate without companies. Sometimes these community activities lead to useable solutions that can be freely used, such as open source software that competes with proprietary corporate solutions, such as Open Word versus Microsoft Word.

Open source communities reveal the source code of their scripts to a nondefinite group of actors. Basically, this constitutes a noncontractual transfer of knowledge without expectation of immediate compensation, although open license agreements might establish rules regarding the

(*Continued*)

(Continued)

extent and the way knowledge is shared and can be reused. Contributors to open source communities are aware that their scripts are part of the joint development of integrated, interrelated designs. While the group codevelops these designs, different participants can exploit the designs differently. Usually, not only private users contribute scripts to open source communities, but commercial contributors also do. Some contributors might use the results privately; others integrate elements of them into products and bundle them into attractive market offerings. For example, IBM and other multinationals use LINUX scripts in their proprietary software (Hertel *et al.*, 2003; Bagozzi and Dholakia, 2006; von Krogh and von Hippel, 2006).

Open design communities focus on the codevelopment of physical objects. Although the software component is also often predominant in open design, because hardware has increasingly become data-centric, the difference is that, at a certain point, tangible products require physical production. Open design thus requires production capacities than can either reside with a producer or can be transferred to the consumer via 3D printing. Examples of codesign platforms range from Thingverse, a MakerBot run platform for sharing user-generated 3D designs, to OpenMoko, an initiative to design a Linux-based Smartphone, to OSGV, a platform initiated by the nonprofit Society for Sustainable Mobility that strives to codevelop a sustainable seven-passenger SUV (Raasch *et al.*, 2009; Kyriakou *et al.*, 2017).

When they started, committed users drove open source and open design communities. These users cherished the freedom from producers that the communities gave them and felt like revolutionaries rebelling against established companies. Conversely, established producers feared the destruction of their revenue. In the interim, producers and users have found ways to close ranks. Various business models developed with varying degrees of closed and open elements allow producers to capture value without alienating the open source/design community members. In turn, users benefit from using parts or all of the open source/design solutions and customizing them to meet their needs. Research on the patterns of open source/design community development, on the range of business models applied in contributions, and on the rights of the users and the producers, as well as on the ways in which value creation and capture occur, is still in its infancy and will increasingly improve our understanding of these phenomena.

2.3 Direct offline tools for opening the solution space

Using direct offline tools, or non-ICT tools, to integrate users as solution information providers comprises different forms of group sessions (e.g., lead user workshops, focus group workshops, customer conferences) and individual creativity sessions (e.g., brainwriting exercises). Although individual sessions have repeatedly proven to provide more creative ideas, innovation practice favors group sessions (Mullen *et al.*, 1991; Litchfield, 2008; Schirr, 2012). Customers can also be integrated into creativity workshops to brainstorm and refine ideas, as well as transform them into concepts.

In contrast to direct online tools, direct offline tools cannot include unlimited numbers of participants, but individual workshop sessions are restricted to quite a small number, often only 15 potential or current customers. Such workshops can be limited to 2-h brainstorming sessions but can also extend over several days. The success of a creativity workshop with customers depends largely on selecting the right participants, on a professional moderator to enable free interaction between the participants, and a good selection of creative and evaluative tools to stimulate and engage the participants (Patton, 2002; Schweitzer *et al.*, 2012).

2.3.1 *Creativity workshops with ordinary users*

Creativity workshops with ordinary users are those that companies organize with a small number of ordinary users, usually between seven and 12, to generate ideas for novel market offerings. Ordinary users may be a company's existing current customers or potential future ones. In such workshops, ordinary users are presented with a task, problem, or question. Creativity techniques are applied to develop ideas that help carry out the task, solve the problem, or answer the question (Jungk and Müller, 1987; Stanfield, 2002). The use of creativity techniques increases the originality and novelty of the ideas that ordinary users generate (Kristensson *et al.*, 2002). Companies' aim with such workshops is to gather original input that reflects consumer needs and serves as an inspiration for product improvement, potential future offerings, and strategic arenas.

In these workshops, a moderator guides users through a creativity session involving the application of creativity techniques. Creativity techniques either take an intuitive–creative approach or a semantic–analytical approach. The intuitive–creative approach follows the principles of free association, analogy building, and intuitive confrontation to encourage unreflective, spontaneous ideas. Brainstorming, brainwriting, the analysis of stimulus words, and the headstand technique are typical techniques within this category. The semantic–analytical approach focuses on the systematic generation of solutions and includes techniques such as morphological analyses, attribute listing, or progressive abstraction. While the number of techniques is constantly increasing, new methods are often combinations and adaptations of the established methods and all methods share certain common elements regarding stimulating participants' creativity. These elements include the transfer of a problem to a different field (analogy), the partitioning of a problem into subproblems, picking up other participants' ideas to build on them, and combining a problem with dissimilar elements (Vahs and Brehm, 2013; Gaubinger *et al.*, 2015). Creativity workshops can either focus on one of these techniques or use them in combination.

Carrying out a creativity workshop involves three phases (Figure 19): preparation, execution, and analysis. In the **preparatory phase**, the innovating company needs to define the topic, classify the type of expected ideas (improvement ideas for current products versus ideas for areas into which future activities could stretch), select and invite users, think about their motivation and rewards, fix administrative details (time, duration, place, required materials, and catering), plan the course of action and creativity techniques used in the workshop, choose and brief a moderator, choose a procedure for documenting the workshop and its results, consider the need for nondisclosure agreements and questionnaires about the users' background.

In the **execution phase**, the actual workshop takes place. At the start of the workshop, the moderator explains the topic, all the participants introduce themselves, the workshop rules are presented, and the expectations are shared. Some time is usually invested upfront in the analysis and formulation of the problem and in the explanation of the creativity technique. If the workshop's focus is on highlighting

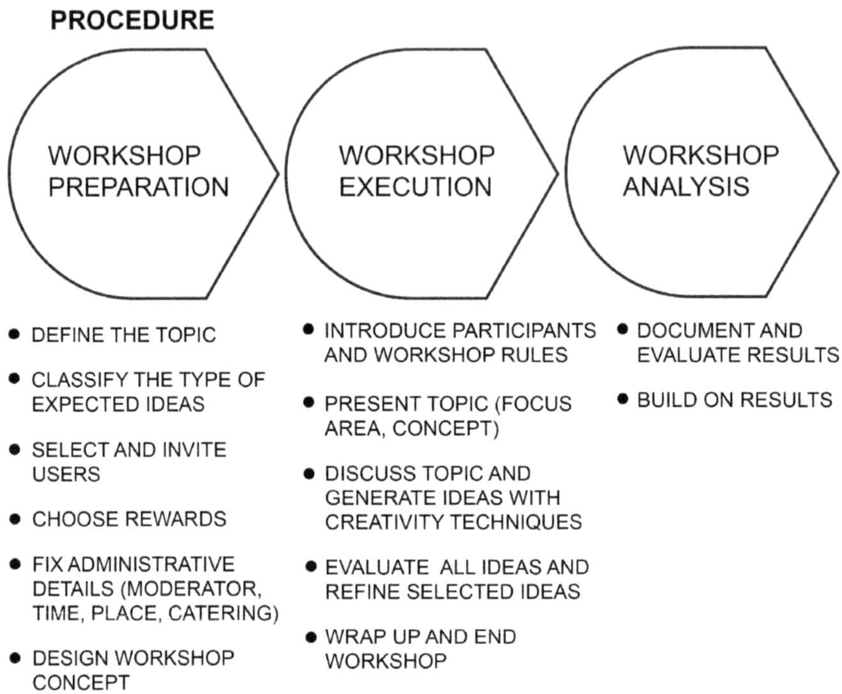

Figure 19: The three phases of creativity workshops.

potential customer problems, with current solutions as a basis for improving them, the start of the workshop also includes gathering the participants' narratives of their positive and negative experiences with company offerings. These narratives serve to identify problems that build interesting starting points for improved offerings. If many issues unfold, they can be grouped into categories that build central topics for the next steps in the workshop.

As a next step in the execution phase, the workshop participants generate alternative ideas, using creativity techniques and with the moderator acting as coach and enabler. If idea generation takes place in smaller subgroups, then each subgroup's ideas are presented after this phase. This is followed by some sort of evaluation process in which ideas are prioritized. Ideas considered more relevant are then forwarded to a new creativity phase in which the participants adapt and refine them to produce more concrete offerings. If a workshop runs more than one creativity phase, the creativity techniques for the different phases should be varied to maintain the participants' enjoyment and engagement. A possible approach could be to first generate a high quantity of ideas through an intuitive–creative technique and to then use a semantic–analytical technique for idea refinement. During the creativity runs, the moderator must stimulate discussion, focus the participants' attention on the central theme, mediate conflict situations, and ensure adherence to the schedule. Depending on the workshop's objective, the moderator should either encourage the participants to imagine and express utopian and unconventional ideas for future environments that differ largely from the existing ones or should restrict their creative activities to realistic ideas for improving existing offerings.

At the end of the workshops, the moderator closes the workshop with a short wrap-up of the key results, a feedback round, an expression of gratitude for participation, and an outlook on the way in which the company plans to use the results (Geschka, 1986; Vahs and Brehm, 2013).

In the **analysis phase** after the workshop, the innovating company documents the workshop, discusses the generated ideas and insights, critically evaluates them, and feeds the relevant ones into the NPD/NSD process. When evaluating and selecting ideas, those doing so should keep in mind that the workshop's goal is central. If the goal was to discover radical ideas, incremental ones should be sifted out and vice versa. This sounds logical, but empirical evidence indicates a tendency to always abandon the more radical ones (see Levinthal and March (1993), Piezunka and Dahlander (2015), and Landau *et al.* (2016) for interesting insights). Informing the participants of the outcome of this analysis process helps motivate them to participate in other customer integration activities for the same project or other innovation projects. We discuss the difficulties that can arise in this phase and the ways to overcome them in Chapter 6.

Customer integration in action: Overview of creativity methods

Brainstorming. Brainstorming is an established creativity technique and allows for gathering a large number of spontaneous ideas quickly. Brainstorming encourages participants to provide any idea and to gain inspiration from the other participants' contributions. In order to assure contributions from everyone, it is important to prohibit criticism of any idea throughout the process.

Method 635. This method is a brainwriting method for groups in which six participants have 5 minutes to write down three ideas. Thereafter, they pass their piece of paper to another participant, read the received ideas as inspiration, and extend or build on them within the next 5 minutes. The procedure is repeated for five rounds.

Blender. The blender aims at integrating the benefits of single solutions into a final solution acceptable to all the participants. Through brainwriting and brainstorming exercises, participants first develop problem solutions individually and then present them. After two participants have presented their ideas, all the participants discuss a way to integrate these two. The next participant presents his or her idea, and integration is again discussed. This multistage process is continued until all the participants have presented their ideas and have found a way to integrate the solutions. The strength of this method is the in-depth, positive discussion of ideas with the potential for holistic, mature solutions. On the downside, it takes time, the participants may perceive this as tiresome, and it may lead to the integration of too many product features, which could result in overengineering.

Tick Tock. Tick Tock is also known as the headstand technique and involves thinking about fears and risks or a market offering that nobody will willingly buy (e.g., holidays in a war zone, holidays for alcoholics). In a first step, the moderator writes down the participants' wildest fears regarding a topic. In a second step, the moderator tries to convert these fears into positives (e.g., holidays to historically relevant places; holidays to study the production, culture, and history of wine) or discusses what the opposite could be (e.g., holidays to peaceful, remote places; holidays for health-conscious travelers). Tick Tock aims to make negative thoughts recede and to allow creativity to flourish.

Osborn technique. This established technique strives to expand on ideas generated via brainstorming. It encourages new solutions through a catalog

(*Continued*)

(*Continued*)

of questions to take the first ideas further. The questions focus on substitution (What might be used/done instead?), combination (Which ideas can be combined to create new ones?), adaptation (What could be changed in the current idea?), modification (How can the idea be altered? Can something be increased/decreased/added?), change in use (Can the idea be used for different ends?), elimination (What can be deleted?), and reversion (What is the opposite?).

Morphological box. The morphological box is a systematic approach to generating ideas within a specific search field by dividing a problem into individual parameters. The participants generate a comprehensive list of possible options for these parameters and thereafter combinations of the different parameters' options.

Table 3 provides an overview of different methods' expected results. Since workshops with ordinary users normally focus on generating first

Table 3: Selected creativity techniques adapted and extended (based on Geschka (1986) and Gaubinger *et al.* (2015)).

Method	Classification	Degree of elaboration of solution	Suitability for creativity workshops with users
Brainstorming	Intuitive–creative	First product ideas, detection of new search fields	With ordinary users With expert users
Method 635	Intuitive–creative	First product ideas, detection of new search fields	With ordinary users With expert users
Blender	Intuitive–creative	Rather mature, complex solutions	With expert users
Tick Tock	Intuitive–creative	First product ideas, detection of new search fields	With ordinary users With expert users
Osborn checklist	Semantic–analytical	Concrete product improvement suggestions	With expert users
Morphological box	Semantic–analytical	Rather mature, complex solutions	With expert users

(*Continued*)

(*Continued*)

impulses for NPD/NSD, intuitive–creative methods are more suitable. Semantic–analytical methods aim at solving complex technical problems and are more appropriate for lead user workshops or workshops with other expert users.

Sources: Michalko (2006) and Gaubinger *et al.* (2015).

Creativity workshops profit from associative chains, which refer to the participants assimilating other participants' ideas and developing them further. Furthermore, such workshops encourage spontaneous, intuitive reactions, as well as systematic problem-solving approaches that extend creative output. Nevertheless, literature reviews of individuals' creative output versus group creativity sessions consistently demonstrate individuals' superiority versus group ideation (Mullen *et al.*, 1991; Schirr, 2012).

One of the reasons for group ideation's inferiority compared to individual ideation is that groups tend to zero in on an idea too quickly. Ways to reduce this problem include asking individuals to generate ideas individually prior to coming to the workshop, to bring them along and present them in the workshop. Another way is to include individual brainstorming sessions, or small subgroup brainstorming, in the workshop. Furthermore, breaks between idea generation sessions improve the idea selection and refinement processes. Popular wisdom says that you should think on a difficult problem for a night before making a decision. Prior research has found much truth in this wisdom. When their intensive thinking of ways to solve a problem leads to the participants being stuck, a break during which they deal with very different topics, or simply relax, fuels their creativity and improves their ability to differentiate between good and bad ideas. Research in the activation spreading field has found that different knowledge domains, which are separated during conscious thinking processes, are activated and linked to one another during such breaks (Anderson, 1983; Greifeneder *et al.*, 2011; Ritter, 2012).

Another way to boost creativity in creativity workshops with ordinary users lies in the choice of participants. More heterogeneous participants are in a better position to generate diverse and novel ideas than more homogeneous ones, because heterogeneous participants stimulate the forced fitting of unrelated issues and the exchange of divergent knowledge and opinions. For example, a mixture of current and potential users is valuable, as the former contribute usage experience with the domain and the latter improvise, draw from their experience in other domains, and broaden the solution space by contributing unusual knowledge. In these mixed settings, potential users receive domain information from current users, thus allowing the potential user to become familiar with the domain. Simultaneously, the current users' exchanges with the potential users challenge these users creatively, entice them to reflect on their habits, and encourage them to change their perspectives. The collective cognition and assimilation processes of these two types of users allows more innovative and useful ideas to be generated than the ideas that homogenous groups of either current or potential customers generate (Hewing, 2013). The choice of participants and setting needs to suits the workshop's aim. Workshops that involve current ordinary users and ask them to reflect on the current solutions and improvement possibilities will generate incrementally new product ideas. If the aim is to generate more radical ideas for existing and new markets, it might be more rewarding to include heterogeneous groups of potential future users and to encourage them to visualize possible futures and to generate ideas based on these futures.

The traditional workshop concept can also be amended to integrate not only ordinary users but also employees, especially those from R&D and marketing, as well as other stakeholders (Jungk and Müller, 1987). Furthermore, including experts from other industries could generate novel solutions, as these experts contribute technological knowledge and other solution information from their industry. Transferring solutions that work in one industry and applying them in another is a relatively low-cost and low-risk opportunity to develop solutions to challenges in one's industry (Enkel and Bader, 2016). Generating and refining ideas together, as well as face to face, enable mutual learning and allow the idea providers to quickly test whether the potential target group values their ideas. In addition, this situation allows all of the participants to jointly design alternatives (Lee and Walsh, 2016). Nevertheless, the inclusion of stakeholders beyond ordinary users also has its downside. Although moderators in creativity workshop are skilled at ensuring that all the participants and their contributions are treated equally and with respect, there are, in real life, differences between their education, experience, social status, nationality, and gender. Their expectations also differ; while ordinary users might participate out of curiosity, or for enjoyment, the innovating company's employees wish to see results and — depending on their personality — might drive this home during the session. The more diverse the workshop settings, the more likely creativity-destroying conflicts are to arise. Besides proper moderation, information about the participants prior to such workshops and the choice of participants with an open-minded attitude are important.

Customer integration in action: The SuperGroup approach

Kantar TNS, a leading market research agency, successfully carries out SuperGroup™ Workshops. These workshops are interactive events, lasting two or several days, during which customers, creative minds, and an interdisciplinary team of Kantar TNS's client generate ideas and refine them through several evaluation and optimization cycles in up to 12 concepts. Kantar TNS has used this method, for example, for innovations in the area of over-the-counter (OTC) products (Figure 20) to develop and position innovative, consumer-oriented applications by involving patients, medical practitioners, pharmacists, as well as representatives from the creative industries and from the client pharmaceutical firm (Kantar TNS, 2017).

Figure 20: Mix of ordinary users and other workshop participants in the workshop session for an OTC product.

The inclusion of more participants through parallel sessions, or online elements, and the choice of unconventional settings are other workshop elements aimed at generating creativity. Creativity workshops include online, or virtual, elements to enable distant searches by inviting customers from around the world to share their ideas in online brainstorming sessions that form part of the workshop. In the same vein, a company can carry out parallel workshops, with each group presenting their creativity task's findings to the other groups before proceeding with another creativity task (Jungk and Müller, 1987). Unconventional settings strive to unleash intuitive and divergent thinking processes and to encourage enjoyment and engagement. With these goals in mind, gamification includes introducing game design elements into nongame contexts. The ability of serious plays to simulate reality, to allow role-playing, and to support collaboration makes such games interesting in terms of creativity sessions. Such games have the potential to produce a climate of confidence that facilitates learning and involvement in new contexts. Agogué and colleagues (2015) used serious games in a workshop with Nutriset, a French SME specialized in treatments for child malnutrition, to develop strategies for new geographical markets. The serious game environment helped the participants visualize the situation in the new market by slipping into the roles of specific personas. Each participant then tried to think and act like the persona assigned to him or her. This procedure encouraged the participants to develop ideas outside the usual firm perspective.

Customer integration in action: Shared steel visions

voestalpine Stahl, an Austrian manufacturer of high-quality steel, carried out a 3-day future workshop with a current corporate customer's interdisciplinary team and voestalpine employees to identify new strategic arenas for innovation. The workshop helped the participants gain an understanding of the key trends affecting the customer's industry and explore strategic arenas for innovations that could help the customer navigate the future. In addition, the immediate interaction between the internals and the externals fostered the customer relationship, their mutual understanding, and trust, resulting in a shared vision for the future and commitment to its realization (Schweitzer, 2013; Rau *et al.*, 2014).

(*Continued*)

(*Continued*)

This case illustrates the role of creativity workshops not only as an innovation tool but also as a marketing tool. As Figure 21 outlines, understanding and using such workshops for both ends enhance the value-creating elements of customer integration into the innovation process.

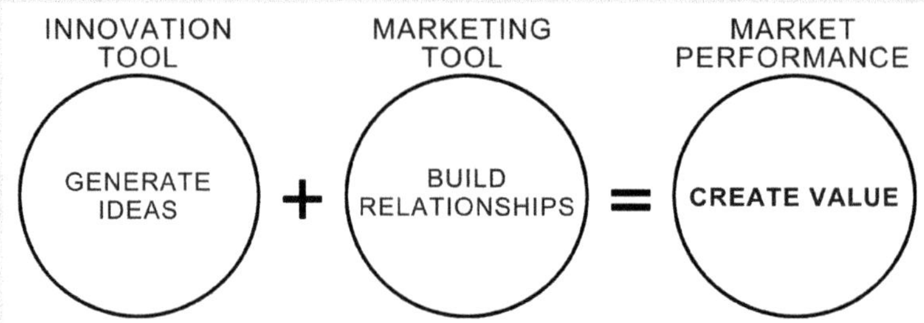

Figure 21: The value of customer integration into interactive creativity workshops.

Table 4 captures the key strengths and challenges that have to be taken into account when carrying out creativity workshops with ordinary users.

Table 4: Strengths and challenges of creativity workshops.

Strengths	Challenges
• Allows for identifying problems that future solutions may strive to solve • Encourages respondents' contemplation of the personal meaning that products, services, and their attributes have for them, their preferences for these, and thus, the generation of needs-based new product ideas • B2C participants are usually easier to recruit for group sessions than for individual creativity sessions • One participant's idea dynamically inspires other participants' ideas	• The risk of biased idea evaluations and too early focus on specific ideas at the detriment of other ideas • The creative group output is lower than in individual brainstorming sessions • Bias or underperformance due to inappropriate participant selection

2.3.2 *Lead user workshops*

Lead users are ahead of a relevant trend and expect to benefit from innovating commercial solutions that are currently available at the marketplace but do not appropriately satisfy needs related to this trend (von Hippel, 1986; Schreier and Prügl, 2008). These users' dissatisfaction with available commercial solutions drives them to experiment and tinker to change these or develop new ones from scratch that better meet their needs. Consequently, they often have a pronounced knowledge of products, components, and technologies (von Hippel, 1986; Urban and von Hippel, 1988; Lilien *et al.*, 2002).

The extent of lead users' innovative activities differs between industries, with high percentages in the medical equipment industry (Lettl *et al.*, 2006a; Chatterji and Fabrizio, 2013), software (Morrison *et al.*, 2000; Franke and von Hippel, 2003), as well as outdoor and extreme sporting equipment areas, such as snowboarding, kitesurfing, mountain-biking, and skateboarding (Franke and Shah, 2003; Lüthje, 2004; Lüthje *et al.*, 2005; Schreier and Prügl, 2008). Lead users feel an intrinsic need for a solution in a field in which few other users currently require solutions. Since their upfront estimate of the potential benefit of having their need satisfied is high, lead users are ready to invest in activities that help provide these benefits. They are willing to innovate on their own, or in cooperation with manufacturers. Provided that the trend becomes relevant for a broader part of the market, their needs also become relevant for other customers in this market. Lead users are months, or even years, ahead of ordinary users in respect of a trend and take a trailblazer position. Consequently, early cooperation with lead users allows manufacturers to profit from their domain-specific skills.

Beside the two core characteristics (ahead of a trend, the expectation of benefitting strongly from an innovation), lead users also often have a strong internal locus of control. They believe that they have the power to exert control over the outcome of life's events and can actively influence their environment. They are often highly involved in a certain product category and tend to develop and test their own prototypes. Their high involvement often makes them opinion leaders in their special area (Schreier and Prügl, 2008), although they are positioned slightly differently in their social networks than typical opinion leaders. The latter have strong direct links with their local network cluster, while lead users have many links between various network clusters (Kratzer and Lettl, 2009). They differ from early adopters (i.e., customers who are among the first to buy a new product) and are not necessarily intensive users of current offerings. There are, in fact, three types of lead users: lead users in the target application and market, i.e., lead users using current solutions in the marketplace, often adopting new products faster and more intensively than ordinary users; lead users of similar applications in analog markets, i.e., lead users who use products in more advanced markets with similar trends and problems; and lead users with respect to product attributes, i.e., lead users who have specific knowledge regarding specific attributes that a future product in the target market requires. For the development of a new X-ray apparatus with higher X-ray resolutions and better mechanisms to detect diseases at an early stage, the first lead user type would be radiologists with strong needs for high-resolution X-rays. The second type could be engineers developing solid state electronic components for microscopic photography. The third type would be musicians or mathematicians with pattern-recognition experience (Schreier and Prügl, 2008; Churchill *et al.*, 2009).

In lead user workshops, lead users and employees interact actively to generate new solutions. As shown in Figure 22, these workshops' process run comprises four phases.

The tasks in the project preparation phase include selecting an interdisciplinary core project team of three to four members, securing top management support for the project, defining strategic arenas and clear goals for the project to help specify the product category, the degree of innovation, and the expected target market. The team members need to develop a familiarity with the topic at an early stage to ensure an accurate strategic arena and goal setting.

In the second phase, the team members identify trends by doing a more thorough search of published material, such as reports on trends in technology, society, and markets, and by interviewing experts and persons

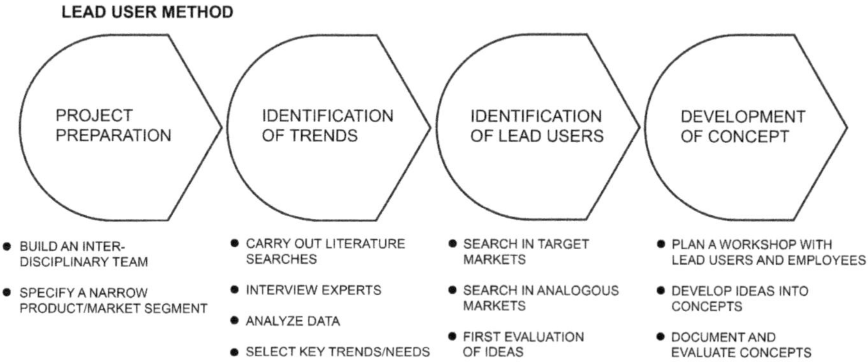

Figure 22: The lead user process. (Adapted from Churchill *et al.* (2009) (Copyright © 2009 MIT Press) and Thomke and Nimgade (1996) (Copyright © 1996 McGraw-Hill).)

with a sound knowledge of the relevant industries and markets. At the end of this phase, the team members should have a broad knowledge of the market needs and technological developments, as well as a list of criteria describing the lead users' characteristics.

The third phase focuses on lead user identification. In this phase, the challenge is to identify users with the desired set of characteristics. Three complexity drivers make the systematic identification of lead users challenging. First, lead users might not be part of the company's target or home market; they could be the competitors' customers or users of products from other industries. Second, lead users do not necessarily have to match the full list of criteria to qualify as workshop participants. Third, some lead users may no longer feel the urge to find a solution to satisfy their need, because they have already developed a solution that satisfies them. Since the problem has already been solved for them, they might be difficult to find, as they are no longer looking for solutions. The most common methods for identifying lead users are screening, pyramiding, and netnography. None of these methods is perfect, but they all have their pros and cons that make them either more or less attractive, depending on a project's specifics. At the end of this stage, the team members interview the identified lead users, discuss their needs and solutions, and generate preliminary concepts.

Customer integration into action: Mastering lead user selection

A number of methods strive to tackle the challenges of the reliable and efficient identification of lead users: screening, pyramiding, broadcasting, and netnography. Figure 23 depicts the key principle of each of these identification methods.

Screening involves surveying a large number of users by means of telephone interviews, or written questionnaires, to identify those who provide answers that qualify them as having lead user characteristics (Franke and Shah, 2003; Lüthje, 2004; Morrison *et al.*, 2004; Prügl and Schreier, 2006). A large number of people's data are checked against the list of criteria that qualify lead users. The voluntary disclosure of personal details is the basis of the selection. For example, if the project team is interested in pattern recognition procedures, a selection criterion for lead users is the importance of improved pattern recognition and own activities to improve current pattern recognition systems. A screening question would then be "How much extra would you pay for X with a Y% better pattern recognition?" and "How many hours have you spent in the last 2 y on trying to improve current pattern recognition systems?". Respondents with an above the mean willingness to pay and above the mean time spent on improving systems should then represent lead users with respect to this trend. Screening is among the more laborious processes to identify lead users and is only economical when the population used to search for lead users is relatively small.

Pyramiding is a specific type of snowball sampling that uses social networks to identify persons with rare traits by sequentially and gradually approaching the right person. Mountaineering is a metaphor for this approach: The stepwise approach to the target person by means of interviews is the mountaineering stage. Pyramiding builds on referrals in such a way that

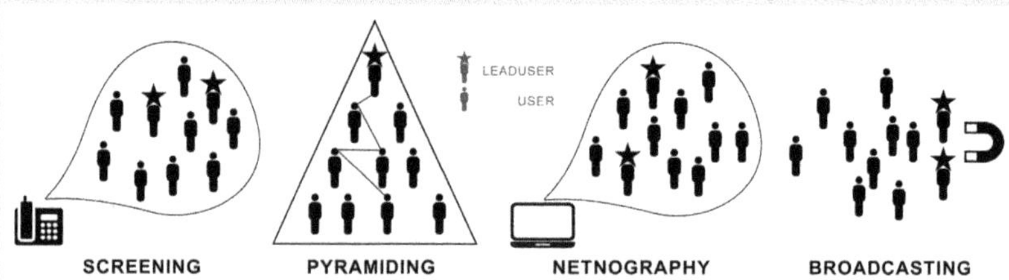

Figure 23: Overview of different lead user selection methods.

(*Continued*)

(*Continued*)

experts within a defined population are asked to recommend other experts who are their seniors in a specific area of expertise. The mentioned expert is then again asked to provide an even more knowledgeable expert. This process is iterated until the innovating company identifies the top experts with the desired lead user characteristics. The identification of the lead user represents the pyramid's pinnacle. The project team starts with an initial catalogue of questions to identify and interview the first experts. The benefit of this method is that project members can adapt and improve their question catalogue throughout their sampling to integrate knowledge generated by their interaction with the interviewees. In addition, this method provides an opportunity to overcome local search bias, because the interviewees may recommend experts outside the technology fields that the project team defined upfront. Three referral levels are often sufficient to identify appropriate lead users (von Hippel *et al.*, 2009; Poetz and Prügl, 2010). While this method is more time saving and cost efficient than screening, it only works well in areas in which experts interact in tight social networks and also depends on a good choice of initial referrers (Bilgram *et al.*, 2008; von Hippel *et al.*, 2009).

Netnography is an ethnographic approach to researching online posted content. This approach, described in detail in Chapter 3, is usually used to identify latent needs by studying customers' interactions in virtual communities. When using netnography for lead user identification, companies study bloggers and their posted ideas and suggestions to identify submitters with lead user attribute. For example, the innovating company can argue that users are highly involved in a theme from the number and quality of their postings related to the theme. Negative postings and complaints indicate dissatisfaction with the current status quo and can allude to a user's lead userness. A user's uploads of prototypes also support lead user characteristics. Nethnography has been successfully used to identify lead users for technologically complex products in advanced analog fields. However, this approach has its challenges for trends that are difficult to limit to a certain technological field and product category and requires scrolling through huge amount of posted content. While trends that are discussed at length call for significant resources, potential users might not discuss others online at all; the value of netnography for lead user identification is

(*Continued*)

(Continued)

therefore dependent on the specific industry, product category, and trend (Belz and Baumbach, 2010).

When using *broadcasting*, a company runs an online idea competition (see Section 2.21) or a virtual stock market for ideas, that is, a playful way of predicting the success of a set of new product options by having, in a stock market game, the participants trade in these new product options' stocks (Spann *et al.*, 2009). Broadcasting appeals to lead users' intrinsic motivation to find a useful solution and relies on a self-selection effect, while all the previously mentioned methods rely on the project team's active search efforts (Hienerth, 2006; Spann *et al.*, 2009). In addition to self-selection, broadcasting allows companies to select lead users based on the content and quality of their submitted material. The method contributes to the effective identification of lead users who are motivated to participate and is particularly useful for attracting lead users from analogous markets. A downside is that it requires the company to make a considerable effort, as it has to develop and conduct the idea competition, or virtual stock market, as well as evaluate contributions and reward the winning contributors (Toubia, 2006; Piller and Walcher, 2006; Bilgram *et al.*, 2008).

Project teams can combine these methods to improve their results, for example, by first using netnographic procedures to identify potential lead users, followed by screening. In screening, qualitative interviews are carried out with potential lead users to gather additional insights into their attitudes, opinion, skills, and reasons for posting content, as well as to help with the fine-grained selection of those best qualified to participate in a workshop. These interviews can also form the basis to build a pyramiding approach (Eisenberg, 2011).

In the fourth and final phase, the project team prepares a workshop, invites the lead users, carries out the workshop, and analyzes the results. Workshops generally last 2 days and include 15–18 persons, of whom 8–10 are lead users and the rest are employees. The workshop starts with an introductory phase during which the moderator establishes a common work context. Thereafter the participants generate ideas and build on these. In this phase, complex tasks can be decomposed into subproblems, with separate sessions generating solutions to these. After an initial generation of solutions, the participants exchange solution information, discuss the findings, and improve and refine them by turning the idea solutions into concepts. Finally, the participants combine and select specific concepts (von Hippel, 1986; Churchill *et al.*, 2009). After the workshop, the project team rediscusses the developed concepts, selects the most promising ones, and moves them into the next corporate NPD/NSD phase.

Conducting lead users workshops allows companies to generate novel solutions with a higher sales potential (von Hippel, 1986; von Hippel *et al.*, 2000; Franke *et al.*, 2006). This approach's advantage over creativity workshops with ordinary customers is that companies do not merely tap the customers' use experience and needs, or ideas for new product areas, but also learn about possible solutions. Companies thus explore lead users' domain-specific skills regarding tricky technical challenges and absorb their knowledge through intensive exchange, while codeveloping concepts. However, identifying lead users is a time-consuming, resource-intensive, and complex task (von Hippel, 2005). In addition, the generated concepts might solve a particular lead user's problem but lack relevance in the mainstream market (Littler *et al.*, 1995; Brockhoff, 2003; Prahalad and Ramaswamy, 2004; Gassmann *et al.*, 2010; Antorini and Muñiz, 2013).

Customer integration in action: Method mix

The core idea of lead user workshops has been amended with regard to other types of workshop concepts where companies, together with expert users, cocreate products from the first ideas to the concepts. One such workshop is the Hackathon (Figure 24). This method is a blend of a workshop with expert users, such as lead users, and hackathons, a word coined by combining "hacker" and "marathon." Hackathons originated in the software industry and describe competitions lasting one to many days, and involving a vibrant mix of participants, ranging from developers, start-ups, students to company representatives. The aim of hackathons is to locate new strategic arenas and quickly transform first ideas in to early concepts. The participating experts often experience such events as entertaining and inspiring, while companies cherish their access to rewarding new ideas. In Hackathons, participants

(*Continued*)

(Continued)

Figure 24: AT&T mobile app hackathon.

Source: https://attmobileapphackathon-hawaii.splashthat.com/ (accessed 30 August 2017).

often work in small groups and are encouraged to exchange ideas with others, with the winner sometimes receiving a considerable amount in prize money. The USD 1 million payout that Salesforce received at the TechCrunch Disrupt conference 2013 was one of the highest prizes ever.

In the early days, hackathons focused on exploratory programming to push the boundaries of software development and were sponsored by leading software companies, such as Sun Microsystems, Hewlett-Packard, and Facebook. Nowadays, hackathons are used for innovation tasks in many industries. For example, Palfinger, a leading Austrian manufacturer of hydraulic lifting and carriage handling solutions, ran a hackathon to find innovative virtual reality, service, and autonomous machine handling. Phillips, a Dutch-headquartered technology multinational, carried out hackathons to find solutions to societal challenges such as innovations that transform cities while simultaneously preserving natural resources. Even governmental entities run hackathons to find innovative ways of improving their services to citizens.

Source: Techcrunch (2013), Johnson and Robinson (2014), Seravalli and Simeone (2016).

Table 5: Strengths and challenges of lead user workshops.

Strengths	Challenges
• Expert users, such as lead users, ideally have usage experience and domain-relevant technical knowledge • Expert users are highly motivated to innovate or participate in innovation tasks. If they regard these innovations as solutions to their personal problems, they perceive participating in the workshop as beneficial for themselves • Access to distant knowledge • Joint workshops of expert users and company employees facilitate the transfer of sticky knowledge • Expert users might already have pioneered product development and might present early prototypes that companies can develop further • Integrating expert users can decrease the costs and increase the speed of problem solving	• If expert users are not representative of a potential attractive future customer group, expert workshops risk developing products for niche markets, or no markets at all • Lead users might only focus on having their problem solved and be reluctant to help solve any other problem • Lead users might behave opportunistically, with consequences such as IP problems, and information leakage • The accurate identification of trends and lead users is resource-intensive preparatory work, with shortcuts in this phase reducing the value of these workshops • The risk of an NIH syndrome if the relevant R&D employees, who have to ensure concepts resulting from a workshop are later realized, are not integrated at an early enough stage

While lead user workshops have proven to be relevant sources for novel concepts, they also have their downsides. The central pros and cons are listed in Table 5.

Chapter 3

Indirect approaches to open the solution space: Methods to identify latent needs

3.1 The relevance of latent needs

Customer needs can be explicit or latent. Explicit needs are current needs of which customers are aware and which they quite actively solicit from the manufacturer. Customers usually articulate such needs as complaints about current product versions and suggestions for improvements they wish to see in future product versions. On the one hand, systematic gathering and analysis of these wants and wishes can engender incremental innovation (Kirca *et al.*, 2005). On the other hand, unsystematic and hectic reactions to these customer requirements run the risk of binding R&D resources to a minor customer base's individual needs satisfaction, which could lead to the served market's tyranny reducing the availability of R&D resources for more systematic and novel innovation endeavors (Hamel and Prahalad, 1994).

A second argument impugning the power of explicit needs satisfaction is related to competitive advantage. Manufacturers innovate successfully if they manage to satisfy unfulfilled needs better than their competition. Since explicit needs and ways to satisfy them are generally well known to all industry players, superior ways of satisfying these only result in product tweaks that do not provide a basis for substantially new products that set off a company's offerings from those of its competitors. Competitors thus quickly follow with similar offerings, and the pioneers do not enjoy long–term, sustained competitive advantage. Latent needs could remedy explicit need satisfaction's shortcoming.

Latent needs are needs that are potentially very important to customers, but they are not (yet) consciously aware of and/or cannot easily articulate them. Figure 25 illustrates this categorization and distinguishes between different forms of latent needs. Latent needs are either hidden, unspoken, or implicit; manufacturers thus find it difficult to identify them (Atuahene-Gima and Ko, 2001; Goffin *et al.*, 2010). While it is easy to uncover overt, explicit needs, latent needs cannot be easily retrieved from customers.

Latent needs are typically not relevant in respect of customers' current product choice, because no product offered currently fulfills these needs. Nevertheless, consumers are delighted if a product offering fulfills such needs, because this makes them feel as if they have always wanted this product. Such products have the potential to provide substantial superior customer value. In other words, products that fulfill latent needs tap into important customer desires and amaze them by anticipating their unarticulated future needs.

Latent needs identification is at the heart of careful, proactive customer orientation (Narver *et al.*, 2004). The identification of latent needs is a rich breeding ground for setting strategic arenas that guide the development of novel offerings. An excellent understanding of the behaviors, longings, and attitudes of potential target groups, as well as the trends and factors that influence the magnitude and direction in which these dimensions evolve is a prerequisite for identifying latent needs.

	CONSCIOUS AWARENESS OF NEED	UNCONSCIOUS AWARENESS OF NEED
NEED IS DIFFICULT TO ARTICULATE	IMPLICIT LATENT NEED	VISCOUS LATENT NEED
NEED IS EASY TO ARTICULATE	EXPLICIT NEED	UNCONSCIOUS LATENT NEED

Figure 25: Categorization of latent needs.

Customers may be conscious of their needs but struggle to convey these accurately to the manufacturer. In this case, the latent need is tacit or implicit (implicit latent need). Codeveloping products with customers, as well as specific ways of questioning (e.g., the repertory grid technique), can remedy this challenge.

Latent needs also include the challenges customers face but have not yet realized (viscous and unconscious latent needs). Providing products that fulfill these needs is likely to surpass customers' expectations and to surprise and delight them. When customers are unconscious of their needs, codeveloping products and specific questioning techniques may fail to provide the desired results. Studying how current products fail in everyday life and detecting unwritten rules and codes, for example, by turning current practices upside down, are ways of retrieving latent needs in such cases. In these cases, immersion into the customers' world via observational approaches (e.g., ethnography) is a way of tapping into latent needs. Techniques to recognize and assess market trends and changes in trends at an early stage are a prerequisite for detecting these needs. Furthermore, the courage to question current customer practices, behaviors, and attitudes, as well as to integrate these with insights into a vision of future markets based on a horizon not confined by current boundaries, helps detect and interpret latent needs accurately (Holt *et al.*, 1984; Reid and de Brentani, 2010).

Customer integration in action: The Unilever way

Unilever, an Anglo-Dutch fast-moving consumer goods company, has a history of running corporate social responsibility (CSR) initiatives and focusing on sustainability issues in its production process. This process focuses on minimizing the company's ecological footprint and optimizing its resource efficiency. Based on insights into sustainability as a latent consumer need, Unilever converts CSR from mere PR initiatives into the core of its business. The company expects consumers to demand and pay for sustainable products if their latent needs are appropriately addressed and satisfied.

(*Continued*)

(*Continued*)

In 2010, with this latent need in mind, the Indian managers of the Lifebuoy soap brand realized that people's everyday lives could be greatly improved if the number of deaths due to infection — which was then 35% in India — could be decreased. Hand-washing would be a cost-effective means of reducing infection and death — a latent need of Indian consumers — because it is a preventative measure against disease, although customers were unaware of this link at that time. In partnership with the government, NGOs (e.g., Oxfam, UNICEF), and local communities, Unilever started an education project, employing local females as educators and distributors of Lifebuoy products. This educational initiative reportedly decreased diarrhea episodes (–25%),

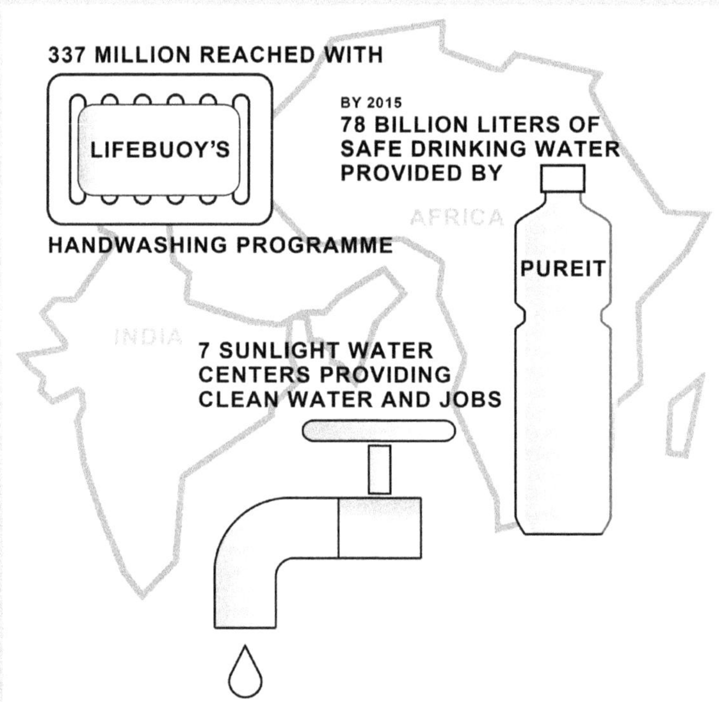

Figure 26: Unilever's collective action for access to water, sanitation, and hygiene (Unilever, 2015, p. 19).

(*Continued*)

(Continued)

acute respiratory infections (–15%), and eye infections (–46%), while simultaneously increasing Unilever's market share (Tideman, 2015).

Unilever extended its engagement in health and hygiene to other countries in Asia and Africa and to other Unilever brands. In 2015, Unilever reached 337 million people with its Lifebuoy health campaigns, provided 55 million people with safe drinking water with Pureit, and in cooperation with partners, built water centers in Africa that provided jobs and access to safe water close to home. At these water centers, Unilever also promote their Sunlight dishwashing, stressing that liquid dishwasher reduces the amount of water required to wash dishes by a third. Local women run the centers, and the proceeds from the sale of water are reinvested in these centers. Unilever also explores models beyond the current business ones by investing in community sanitary projects, agricultural water management, and conservation initiatives (e.g., via the Hindustan Unilever Foundation). Unilever's aim is to have improved the health and hygiene of over a billion people by 2020 (Figure 26). While critics may consider this just a huge marketing initiative to sell more Unilever products, the improvements the company realized in many communities worldwide are impressive (see https://www.unilever.com/sustainable-living (accessed 30 August 2017)).

3.2 Ethnography

Ethnography is a motive- and behavior-oriented long-term research method — originating from cultural anthropology — with an exploratory nature and is used to systematically study people, tribes, cultures, and societies. Market researchers and innovation managers have applied ethnographic methodology to understand how people use consumption objects and connect via them within the context of real-world consumer experiences and how this knowledge can be used for NPD.

In ethnographic research, the researcher acts as a participant and an observer of the studied target group. The researcher engages in daily face-to-face contact with a small number of people from the target group for a prolonged period (several weeks to years). The researcher participates in the target group's life, studies everyday life as it is actually lived, and gathers all sorts of data ranging from written field notes to recorded media, including videos and photos. This immersion captures real insider knowledge and authentic insights into the cultural contexts and social phenomena.

The evidence collection method and analytical processes to capture and analyze ethnographic research is not set, but appropriate, systematic modes of data handling develop while a project unfolds. Researchers organize field notes, transcribe interviews, and repeatedly compare, contrast, and sort the different forms of texts and videos to identify patterns of behavior, values, and beliefs. The analytical process lasts from the beginning of the researcher's fieldwork to post fieldwork interpretations once the researcher has left the research site to make sense of the experience.

For example, Schembri (2009) and Schembri and Boyle (2013) carried out an ethnographic study of the Australian Harley-Davidson Owners Group during fieldwork that lasted more than 3 y, which included observations, interviews, and the capture of several hours of visual documentations. The study finds that this group consists of empty nesters with plenty of money who enjoy the annual rallies that bind them and the feeling of freedom while riding. Three key elements are central to these group members: socializing, communing, and indulging in consumption as play, which includes the customization and ritual jig and polishing of their bikes. While critical persons might sneer at such research as an expensive and time-consuming way of discovering rather banal findings, which are not completely new, the allure of such research lies in the vibrant depictions of environments and lifestyles. The Harley corporation learned about the many ways in which the bikers strive to be symbolically personified by their individualized bikes (Figure 27). Ethnography allows an innovating company to see the world through the target group's eyes and to internalize the intricacies that differentiate cool from awkward. Such a company then becomes accustomed to the specific attitudes, actions, and artifacts of particular social groups expressed through language, looks, and activities, and these form a rich source for understanding hidden needs and for inspiring future product development.

Ethnographic research does not provide immediate hints on the required product features for product innovation and may thus be perceived as remote from the market. If an innovating company expects

Figure 27: Insights into the role of consumption for members of Harley owner groups.

immediate ideas for product innovation based on this technique, it might be disappointed. Instead, ethnography allows for developing a detailed and complex understanding of the current user environment, for extrapolating the evolution of that environment into the future, and for imagining future solutions based on these insights.

Customer integration in action: Anthropologic approaches for studying markets

Ethnographic research provides an invaluable source of authentic insights, notably regarding target groups that build subcultures of enthusiastic product users. Such enthusiastic subcultures are not limited to B2C products; for example, Rosenbauer, a fire trucks manufacturer, profits from ethnographic insights originating from the high percentage of its employees enrolled in voluntary fire brigades. The insights allow the company to gather the distinctive attitudes, actions, rituals, and values of this specific subculture on a continuous basis.

A rich source for further examples on how to use ethnographic research for new product and service improvement is Ellen Isaacs TED video: https://www.youtube.com/watch?v=nV0jY5VgymI (accessed 30 August 2017).

Ethnographic researchers are interwoven with the individuals they study. This entanglement impedes objectivity, and the results depend strongly on the researcher's interests and methodical competence.

Consequently, such researchers have to be aware of bias due to their subjective attitudes, experience, and understanding that could direct their inquiry and analysis.

On the other hand, the observer effect in ethnographic research is lower than in other observational techniques due to the research's extended time period leading to respondent habituation to the researcher's presence. This habituation ensures more natural behavior than in shorter term observations, during which the participants may be more prone to changing their behavior due to their awareness of being observed.

Researchers need to exercise discretion and be sensitive to ethical issues throughout their fieldwork and analysis. A verbal consent script should inform the participants that researchers intend to observe and interview them; those who refuse to give their consent should be omitted. It is particularly pivotal to find open-minded respondents for video data, whose anonymity and confidentiality are difficult to assure, and to build and uphold credibility and mutual trust to obtain rich data.

The aim of adaptations of the traditional ethnographic approach is to gather profound insights into how products and brands fit into consumers' everyday life but with less time effort. Such methods include live-ins (i.e., accompanied shopping sprees, home visits, or site visits that range from a few hours to several days) and the study of (electronic) diaries (e.g., blogs, video diaries) (Table 6).

Table 6: Strengths and challenges of ethnographic methods.

Strengths	Challenges
• Rich qualitative findings due to close, intensive, long-term interaction • Flexibility/adaptability to a variety of circumstances • Rather low observer effect • Authentic insights into the target group's distinctive practices, motives, and values • Insights into situations that customers encounter in a real-life environment	• Entanglement hampers objectivity • Remoteness from the market • Time- and personnel-intensive research • Difficult to analyze due to high volume of data • Unrepresentative small samples

3.3 Empathic design

Empathic design is a less time-consuming approach to ethnographic research than traditional ethnography. Based on the idea that customers may find it difficult to clearly articulate their needs, empathic design focuses on understanding customer needs through empathy with their world. Method wise, the approach focuses on observing behavior and interviewing customers about their reasons for their behavior, rather than interviewing them about their needs. In empathic design, the innovating company specifies the situations, or activities, that it plans to observe. Rather than undertaking longitudinal studies of a few customers' diverse sets of daily activities — which traditional ethnography does — researchers observe a higher number of customers performing a small number of activities. These customers carry out these activities in their natural surroundings and with the devices they usually utilize, because studio settings decrease insights' authenticity. Empathic design uncovers the ways customers have developed to cope with inconvenient situations and which they perform so habitually that they would be unable to articulate the inconvenient situation in an interview. Once a product can solve this inconvenient situation in a practical way that is better than the customers' habitual workaround practices, they welcome the product enthusiastically, because it fulfills a latent need (Leonard and Rayport, 1997; McDonagh-Philp and Lebbon, 2015).

Empathic studies can also provide insights into the ways customers have amended a product, or use products in combination with other products, to better serve their needs. Such studies could also show that customers use products in other than the intended ways, which might identify opportunities for new markets. Moreover, empathic design can uncover emotions toward products or usage situations. Understanding positive and negative emotions can serve as inspiration for new or adapted products. For example, London Underground discovered a gap between the positive results in passenger satisfaction surveys and the negative results in brand image surveys and wanted to understand this gap. It used empathic design in the form of accompanied underground journeys and found that passengers have a deep-seated, unconscious fear of being underground. This fear was largely based on around feelings of being in a dark, dirty, dismal, and potentially dangerous place. Based on these insights, London Underground emphasized that it would create a more welcoming, lighter, and cleaner ambience (Boddy, 2011).

Empathic design involves five steps (Leonard and Rayport, 1997):

1. *research design phase* during which a company defines the observation subjects and situations and chooses the employees to integrate into the empathic design project
2. *data gathering phase* involving visual observation, which ideally includes videotaping to avoid bias, as well as interviewing to facilitate the interpretation of the observations
3. *analysis and reflection phase* during which the information gathered by an analysis team is reviewed
4. *brainstorming for solutions phase* comprising a creative search for solutions to latent needs discovered in the previous steps
5. *prototype development phase* during which the prototypes for solutions are created and presented to customers for early feedback and improvement suggestions.

The downside of this method is that empathic design projects might not live up to their expectations as they might fail to uncover any problems customers encounter in their daily interaction with specific problems. The projects might only discover small problems that are irrelevant for a larger market. Insights into the latent needs of such studies' small number of participants give no indication of their relevance for potential target markets. Further, the vast volume of videotaped data is very difficult to screen, analyze, and interpret. Gathering the data is only one part of the story; the other one is interpretation. The data do not directly indicate the right path to follow, only providing a company with insights into personal user stories and experiences. The way in which the company interprets these data and uses them for later creative processes determines their success. However, the method focuses on the way the data are gathered and provides little guidance for their accurate interpretation and analysis (Postma *et al.*, 2012) (Table 7).

Table 7: Strengths and challenges of empathic design.

Strengths	Challenges
• Rich qualitative findings regarding specific usage situations • Insights into situations customers encounter in a real-life environment	• Risk of not finding novel insights • Difficult to analyze due to the high volume of the data • Unrepresentative small samples

3.4 Netnography

The portmanteau word netnography combines "inter*net*" with "eth*nography*" and is an explorative, mainly text-focused, method that applies ethnographic principles to analyze user-generated online content. Netnography comprises the passive, lurking, unobtrusive observation of a target group on the web. For example, an innovating company can track online conversations and blogs, or view audiovisual content of online community members. Netnography also has an active component that involves participation in online communities through real-time participation in discussions, invitations to interview, or by tracking conversations.

The crux of netnography is the study and analysis of consumer opinions and dialogues derived from blogs, emails, fora, tweets, and online communities. Individuals who actively discuss products online are usually more involved consumers, who might thus not be representative of mainstream customers. However, they are likely to be opinion leaders who could influence mainstream customers. Listening to these customers' discussions of their likes and dislikes not only helps gather explicitly verbalized needs but also provides insights into implicit existing needs. Netnography involves a systematic procedure to gather these needs from online conversations through the following steps (Kozinets, 2006; Füller *et al.*, 2007):

1. determination of user characteristics
2. community identification and selection
3. observation and data gathering
4. data analysis and interpretation of findings.

First, the innovating company has to determine the user characteristics as a prerequisite for searching for online places where these individuals are active. For example, if the goals of the research are gathering insights to improve existing basketball shoes and gaining inspiration for completely new shoe designs, highly enthusiastic basketball players with broad product use expertise are a likely sample for netnography (Füller *et al.*, 2007).

In a second step, the company looks for electronic bulletin boards (e.g., groups.google.com, egroups.com), web rings of thematically linked pages (e.g., www.epinions.com), blogs, and social media groups where individuals with the specified user characteristics exchange information and ideas. The number of online places for a specific area of interest may easily overwhelm an innovating company. For example, basketball-related topics are discussed in more than 500 online communities. Although there is a supporting screening software (e.g., Meltwater, Netbase, SocialHub), a more realistic and effective option than gathering netnographic insights in all these communities is to concentrate on a selection of communities with the highest likelihood of providing valuable insights. Selection criteria for identifying such communities are the number of postings in general, the number of postings focusing on the investigated product category, the message posters' profiles (e.g., Klout-Rating, years of expertise), the number of message posters, the elaboration of the posted content, and the intensity of the activities commenting on the postings. Further, a community may be selected if its posted comments and reviews verifiably influence purchase decisions or if various credible sources mention it as a leading online community (Arnould and Cayla, 2015; Meyer *et al.*, 2016).

In step three, researchers observe activities in the communities, screen posts and store data for further analyses. For example, for the abovementioned study of basketball shoes, specialists at Hyve, a German online innovation consulting agency, chose five communities (e.g., Niketalk, Solecollector, Kickz101) as data sources, gathered and filed data for an extended half-year period, and scrutinized over 240,000 screened posts to explore ideas for product improvement and innovation (Füller *et al.*, 2007).

Customer integration in action: Learning from the customer's customer in the B2B world

B2B companies often draw on insights gathered from their immediate customers to improve their products. Bekaert, a Belgian manufacturer of steel wire transformation and coatings, moved beyond its immediate customers to gather needs information from consumers. Bekaert offers a range of brassiere wires to brassiere wire manufacturers, which in turn sell the shaped wire to lingerie wholesale and retail businesses. Bekaert used netnographic approaches to study end-consumers' social media discussions on bras, learning that the end-consumers were dissatisfied with the wires, which broke, squeaked and ruined the fabric. Bekaert developed a wire that specifically addressed this problem and strengthened its position as an innovative business partner for its immediate customers.

Source: Muylle *et al.* (2016).

The fourth step is asking the permission of the selected online communities' web administrators to analyze their data. Community members discussing topics online do not directly provide researchers with information, which they do when ethnographic approaches and many other market research methods are used. Instead, the research gathers secondary data that community members produce with other goals in mind than providing researchers with data. However, community policies — if so agreed by all the members — often allow the administrator to share the data exchanged in the communities with others upon request. Asking administrators for data is therefore an unobtrusive and effective way of abiding with market research ethics. In addition, researchers need to assure the informants of their data's confidentiality and anonymity and first ask the community members' consent if they use direct quotes (Kozinets, 2002).

As soon as researchers have received permission, they apply qualitative analysis tools (e.g., MAXQDA software) to categorize and interpret the data. They group user comments according to their topics to identify hotly discussed ones, which indicates their popularity and relevance. In addition, the researchers differentiate between complaints about and praise for current products and register user ideas for improving them, as well as suggestions for completely new products.

The Hyve project for basketball footwear identified 24 footwear components, of which the design, cushioning, and comfort were the three liveliest discussed ones. While the majority of discussions provide current model feedback, online community member discussions provide insights into their latent needs, allowing an innovating company to understand why people prefer certain feature designs over others. Community members also share information on how to improve models that do not have desired features. For example, they provide customization guidelines, which not only reflect the importance of individualization for basketball shoe wearers but also hint at technologies that prove effective (e.g., online reactions to the provided advice indicate its effectiveness) in respect of certain customizations, such as applying glitter in a way that ensures the shoe remains flexible and crease resistant (Füller *et al.*, 2007).

Using netnography for the same holistic view of a target group's behavior that ethnography provides is difficult, because the innovating company is rarely given a full picture of this group due to online communication's lack of richness. Triangulation and combination with offline observations would be interesting but is extremely difficult to obtain due to difficulties with informant identity. Users often build artificial representations of the self (Belk, 2014), which do not reflect them in real life (Weiss *et al.*, 2008). However, netnography allows (1) insights into the product aspects that customers discuss, (2) the positive and negative emotions in such discussions, and (3) the ways in which users have improved the current products (Table 8).

3.5 Focus group workshops

Focus group workshops are moderated, interactive discussions between a small number of customers. Such workshops usually include 8–12 participants, one or two moderators steer them, they are semistructured, and last 1–2 h. Focus group workshops can take place online or offline. In online sessions, the number of participants can be higher (up to 30) and the participants can be geographically scattered, but the workshop duration

Table 8: Strengths and challenges of netnographic methods.

Strengths	Weaknesses
• Largely unbiased consumer insights from passive lurking approaches • Plentiful and easy to obtain data that can be accessed unobtrusively • Data on user emotions toward products, as well as on user innovations and improvement suggestions	• Identifying appropriate online groups, as well as filtering and analyzing the relevant topics that they discuss, is time and resource consuming, although some of the tasks can be outsourced to software programs • There are lively discussion groups for some product categories but not for others — netnographic approaches are thus not useful for all product categories • False user identities can compromise insights • Lurking can be perceived as an inappropriate intrusion into customers' private lives

should be shorter than in offline sessions, as the participants are more likely to be distracted, less engaged, or lose interest faster.

Focus group workshops are a relatively fast and cheap way of gathering insights into the feelings and opinions of a spectrum of different customers. It is a very flexible technique, and the workshop concept can be adapted to fit the goals of different NPD/NSD phases: The workshop can either (1) act as a direct approach to open the solution space by including creativity techniques (see Section 2.3.1) or (2) focus on direct approaches through latent needs by including introspective and projective techniques or (3) be a way of testing the qualitative concept by including typical concept test questions (see Chapter 4). In this section, we focus on the second workshop concept.

A moderation guideline needs to be developed to ensuring a clear structure and the discussion's logical flow. This guideline comprises the introduction (explanation of the purpose and the basic principles), the warm-up (the participants introduce themselves and their experience with the topic, these words serve to get the participants talking and make them comfortable), the discussion of the main topics (some key questions and also room for following up new questions that emerge in a specific discussion), and finally, the closing comments and thanks.

The moderator is responsible for keeping the conversation lively and for studying the topic from different angles. The moderator contributes probes and prompts (e.g., What do you think about that? Tell me more about that!), controls dominant participants, and encourages interaction between the participants to allow everyone to comment, discuss, and reflect. If disputes arise, the moderator helps the participants find ways to reconcile and adjust their views.

The moderator is not necessarily an industry expert but has to fit the participant group. For example, a female moderator is more appropriate for feminine topics, while a male moderator is a better choice than a female one for a male group of heavy beer drinkers.

Not only the choice of moderator but also the compilation of the participants needs attention. Group hierarchy is problematic. For example, if a factory worker and a supervisor are in the same group, this hinders an open and candid discussion. More generally, the disclosure of sensitive issues is more embarrassing in heterogeneous than in homogeneous groups. Focus groups with an emphasis on creativity profit from heterogeneous groups that inspire each other through their ideas. Conversely, focus groups aiming at understanding deeper-lying motives and feelings, and whose participants are more homogeneous, are more likely to stimulate discussions and use group effects with one participant's statements triggering others' comments.

The risk of group dynamics is that shy and unconfident participants with perceptions that differ from those of articulated views are reluctant to reveal their true view and merely mimic the dominant persons' response. This way, important facets of the focal topic can be missed. A competent moderator can partially countervail such behavior. Further, it is wise to conduct several focus groups to minimize this problem. If the innovating company has 12 participants, it is better to run three separate focus groups with four participants than a large one with all the participants, as this reduces bias due to dominant personalities.

All in all, a focus group workshop constitutes a relatively uncomplicated, fast technique for capturing a spectrum of customers' opinions, feelings, spontaneous reactions, and attitudes toward a topic. An innovating company can gather vivid insights into the way customers communicate about a topic and their natural way of articulating and verbalizing. Such insights help the company gain an understanding of certain opinions' backdrop, thus providing innovation teams with food for

Customer integration in action: Looking into the hearts of excavator drivers through focus groups

Wacker-Neuson, a leading German manufacturer of high-quality construction equipment and compact construction machines, carried out a focus group with excavator drivers in Austria. This customer integration approach aimed at identifying the explicit and latent needs of these drivers as an inspiration for future NPD. Wacker-Neuson invited eight excavator drivers to a 2-h focus group workshop. After this workshop, Wacker-Neuson invited the participants to lunch and gave them a guided on-site tour.

In the focus group workshop, the participants were invited to talk about their likes and dislikes of the current products and to provide improvement ideas. In this phase of the workshop, the majority of workshop participants' articulated needs pertained to the improvement of their cabin comfort and the optimization of technical features, including the power tilt and attachments. Later, they were prompted to share their experience on the construction site by telling personal stories. In these storytelling sessions, the drivers revealed latent needs in quite different areas, predominately related to safety and communication. They reported accidents due to their limited possibilities of communicating with the foreman and because they could not see and connect with their coworkers. Furthermore, all the participants felt that the drivers' job is a lonely one and indicated that they would not only be happier but also feel safer with more communication possibilities.

They also raised concerns about their lack of keys, their access authorization, and the immobilizer systems that encouraged theft. In the words of one focus group participant: "Any child could start the machine; this is not only dangerous, but it is also incredible considering what such an excavator costs and anyone could come along and take it." In further discussions on the "lack-of-key" topic, the participants revealed that their central concern in respect of this topic was that they feel that their lack of keys diminished their excavators' value. They take pride in their machines and feel that keys and an authorization system would symbolize the value of the machines.

This example shows that the majority of needs information gathered in focus groups is explicitly articulated product improvement information. Identifying latent needs is rare, but once found, they are of great value for more far-reaching product improvements with a higher potential of creating unique product attributes that build on the participants' personal values.

thought. Focus group workshops can also include idea generation elements or concept evaluation tasks. Nevertheless, the innovating company should be clear whether the focus is on direct approaches to open the solution space (gather ideas), on indirect approaches to open the solution space (gather immediate and latent needs), or on approaches to close the solution space (gather concept evaluation information). This clarity is important, because these different approaches demand a different choice of participants, for example, more heterogeneous samples in approaches to open the solution space and more homogeneous samples in the other approaches. Second, they demand different workshop materials and exercises, for example, a detailed concept might be helpful for the latter approach but harmful for the exploration of latent needs. Third, these approaches are generally used at different times in the NPD process; for example, latent needs detection is useful at the very start to set strategic arenas for future NPD projects, while the value of focus groups as a means of conducting qualitative concept tests only materializes clearly at the end of the fuzzy front-end phase and in the development phase of an NPD project.

Focus groups are usually carried out to discuss the characteristics and perceived benefits of existing products or concepts. Revealing latent needs through focus groups is not an easy task and there is no guarantee of success. If the participants have to discuss a specific product, or compare it to others, and focus strictly on actual features, they are more likely to generate data for incremental product improvements. If they more openly discuss their experiences with products, and situations in which they use, or do not use, certain products, focus groups are more likely to deliver latent needs information. For example, in a focus group workshop on ready-made baking mixtures for cakes, housewives first tasted these cakes, then discussed the taste, color, and other apparent characteristics of the different cakes. One of the participants mentioned "Wow, they all taste quite nice. A pity that using them makes you feel so guilty. You just mix them instead of baking them yourself." The others nodded and a whole discussion evolved about self-esteem, cooking for loved ones, and the feeling of achievement with do-it-yourself and pride. Based on these insights, a leading manufacturer of cake mixtures for

Table 9: Strengths and challenges of focus groups.

Strengths	Weaknesses
• Insights into ways of articulating feelings, opinions, and attitudes • Fast and low-cost method to access needs information • New impulses through group effects • The qualitative research approach conveys the complexities and subtleties of specific customers, customer types, and the different ways they perceive the specific topic • B2C participants have more fun and feel more comfortable than with personal interviews • It is versatile technique that can be used for direct/indirect approaches to open/close the solution space	• The ability to capture latent needs hinges on the moderator's psychological abilities and the participants' articulation ability and willingness • The small sample size does not allow quantitative statistical inferences regarding the target group • Respondents with a view that differs largely from that of others might be reluctant to express their needs in the group

consumers adapted its mixtures. Previously, consumers only had to add some water, mix, and put the mixture in the oven. Afterwards, consumers had to add milk, eggs, and butter before mixing and baking. Moreover, the manufacturer started to develop recipe booklets based on the mixtures and suggestion for decorations. It also started online interactions that provide consumers with space to upload and share their cake creations. These changes gave consumers a more intense feeling of self-accomplishment when using baking mixtures, and the manufacturer's turnover soared (Table 9).

3.6 Service blueprinting and customer journey workshops

As the name suggests, this method was originally intended for service innovations, but it can be adapted for product–service hybrids and for physical products' purchasing process.

Service blueprints are visual representations of customer service encounters. They chronologically depict the interaction between the customer and the frontline employee or online service interface. They also describe the consumer and the service provider's roles, including the activities that are visible to the customer and the background, supporting activities that are relevant in terms of delivering the service to the customer.

Service blueprints comprise five components (Zeithaml *et al.*, 2006), which Figure 28 illustrates in terms of an online hotel booking services. The vertical lines between the different actions in a blueprint highlight the connections between the components:

- customer actions
- onstage actions
- backstage actions
- support processes
- physical evidence.

Customer actions are visualized on the top of a service blueprint and chronologically describe the steps a customer experiences when using a service. Customer actions are the crux of service blueprints and the starting point for designing the blueprint's other components.

Onstage actions are those that a frontline employee, or a service touchpoint technology, undertakes that the customer can see (visible actions). If customers perform a certain activity, the interaction line with the onstage employee action is crossed and a moment of truth occurs. During this moment, the service provider interacts with the customers in a way that either satisfies or dissatisfies them and motivates them to either move further down the service encounter steps or discontinue their interaction.

The backstage actions include all those that the customer cannot see (invisible actions) that are required to perform the onstage actions. For example, a customer checking the availability of a hotel room at a certain date by calling the reception (customer action), receives information

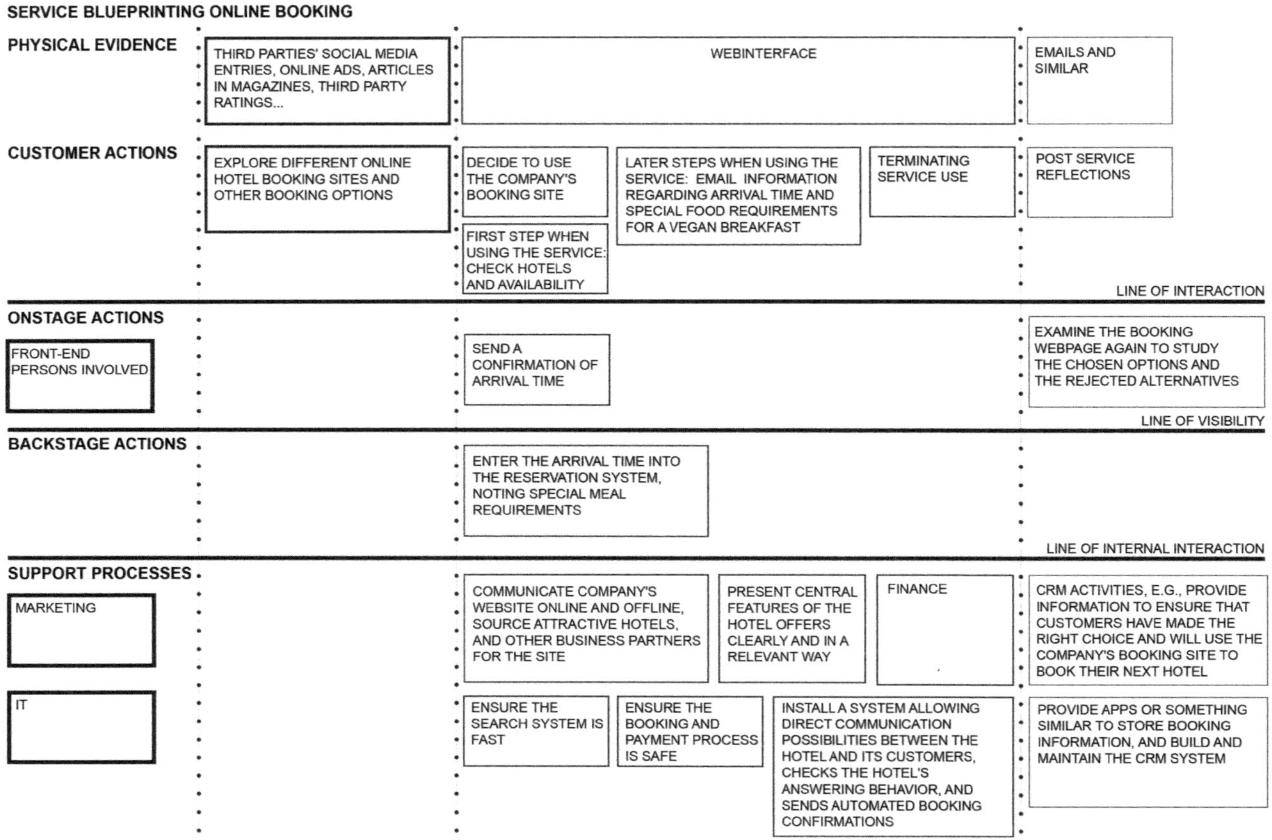

Figure 28: Exhibit of an illustrative service blueprint for an online hotel booking service.

whether a room is available at that date or not (onstage action). The receptionist's backstage actions include entering the date into the booking system to check whether a room is available at the relevant time (backstage action).

Support processes are those that the contact employees do not carry out, but other individuals or units do. In this example, this would be the process of developing, updating, and improving the booking system that the receptionist uses to check room availability.

The physical evidence, along with the customer actions, is positioned at the top of the blueprint and represents the tangibles that a customer faces. These tangibles can influence customers' service quality perceptions.

The blueprint is based on the customers' perspective of the service process and highlights their role as well as the employees' roles. The blueprint's focus is on the internal processes required to realize a specific service by clarifying the human-to-human and the human-to-machine interfaces.

Blueprints can be used for high-level overviews of conceptual processes (e.g., the entire process of a customer's booking of a hotel, beginning with a web search for appropriate hotels to finalizing the booking) to a detailed description of a subprocess (e.g., the credit card payment process). Once a service has been launched, the innovating company can link the blueprints to internal IT-based process tools to involve different units in the different phases of the service realization.

Service blueprints are quite simple, and workshop participants usually find them easy to understand. In service blueprinting workshops, the innovating company jointly develops the initial service blueprints with customers or presents customers with internally created blueprints to identify improvement points and fail points. In these workshops, customers convey their wishes, needs, and attitudes in respect of a certain service, after which the employees and customers discuss how a specific service blueprint corresponds or fails to correspond to these. Such workshops can continue by jointly developing mock or simulated service delivery processes, comparing them with the proposed initial blueprints, and evaluating their attractiveness for potential customers (Alam, 2006).

Before inviting participants to the workshop, the innovating company has to specify the service process or subprocess, the workshop, and the process's customer segment, all of which form the workshop's target. Subsequently, representative customers of the target segment and employees are invited to the workshop. The actual service blueprinting workshop for service innovation can either start with a description of the existing service blueprints or idealized new blueprints. In the former case, blueprints visualize the current processes and activities. Such service blueprints allow for detecting obstacles due to insufficient service planning or poor communication between the different service design levels in the existing services. These insights then serve as a basis for service redesign and improvement. In the latter case, the starting point is an idealized service encounter based on customers' description of an ideal process at the customer action level, which is followed by a discussion of the required onstage and backstage employee actions for the smooth realization of the customer actions. In a further step, the support processes and physical evidence are considered. The workshop closes with suggestions for future innovation based on the blueprints (Bitner *et al.*, 2008).

Customer journey workshops are variations of service blueprinting workshops. Similar to blueprinting workshops, the participants are a mix of employees and customers. In these workshops, a customer journey map is created. This map also delineates the different steps that a customer experiences from the beginning to the end of the service encounter at the top of the map (stages and doing). Instead of then focusing on the internal processes, as in service blueprinting, customer journeys focus on the empathic understanding of the customer throughout the mapped service encounter by gathering insights into the customers' thoughts and feels throughout the different stages. Based on these insights into the customers' perceptions, the workshop participants develop potential opportunities for service improvement and innovation (Table 10). Figure 29 provides a customer journey map, using the same example as in the blueprinting map, for a comparison of the differences and similarities between the two approaches (Kalbach, 2016).

Table 10: Strengths and challenges of service blueprinting and customer journey workshops.

Strengths	Weaknesses
• Captures customer service experience from the customer's point of view • Provides insights into ideal service experience from the customer's point of view • Participants find this method simple and easy to understand • Makes roles and interdependencies transparent • Relatively fast and low-cost method of accessing customer service experience information	• Quality of the outcome depends on the selection of employees and customers with the right mentality and personality to participate productively in the workshop Limited representativeness of the participating customers' experiences with the current processes and developed idealized processes due to the small number of participating customers

3.7 Laddering technique

The laddering technique is a specific way of questioning used in in-depth personal interviews to retrieve the deeper-lying reasons and motives for specific customer behaviors or desires. In-depth interviews are an unstructured, qualitative, one-to-one approach to interviewing that requires highly skilled interviewers and can be very flexibly applied or adapted to specific goals.

The means–end chain model is the theoretical model on which the laddering technique builds. Means–end theories of consumption hold that consumers have certain values and goals in life which influence their behavior. Product or brand attributes, the perceived consequences of a certain behavior, and individuals' moral concepts interact in a way that makes products, brands, or consumption activities the means to reach a certain end. Means–end chain approaches strive to understand the relationship between the meanings certain products have for an individual and this individual's moral concept. These approaches strive to uncover the linkages between "where a person wants to be and the means chosen to get there" (Gutman, 1982, p. 68). According to these approaches, customers who need to decide between different product

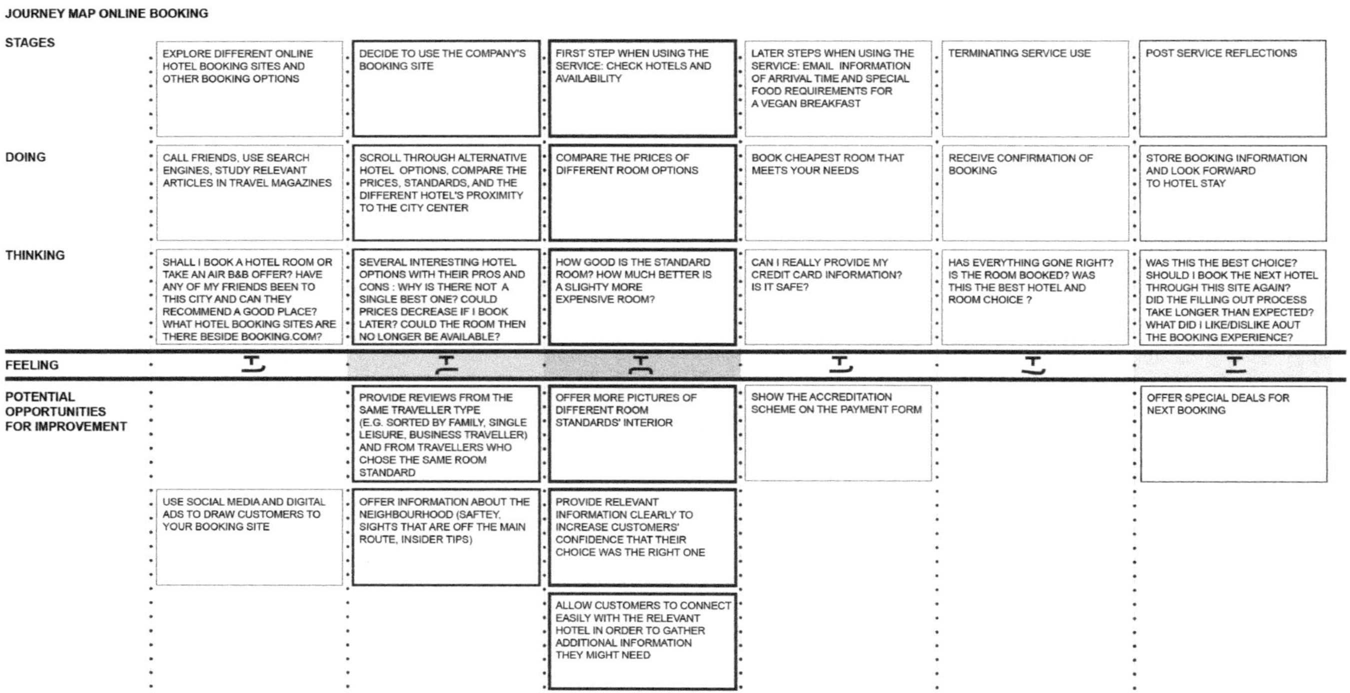

Figure 29: Exhibit of an experience journey map for an online hotel booking service.

alternatives first examine the specific, immediate product features (e.g., one apple is red, the other one is green; one car has a long, flat body, the other is compact) and the automatically associated abstract product features (e.g., the red apple is sweet, the green one is sour; the car with the flat body is racy and fast, the compact one is economical and maneuverable). The association between specific and abstract product features is rooted in the cognitive schemes that individuals construct to navigate a complex world. At the next level of abstraction, customers consider the consequences of product usage. Customers generally ponder what they could gain from consuming or using a product, how they will feel when doing so, and what others would think of them when they do so. Their expectations of the consequences of consuming or using a product can be emotionally laden (e.g., I will feel good and healthy after eating the apple), or function based (e.g., the apple contains many vitamins), and include negative expectations (e.g., the sporty car might not have enough space and have a high fuel consumption) and positive expectations (e.g., I will look sporty and dynamic in this car, driving this car will be fun). The next level comprises customers' moral concept, which builds their lifestyle, core values, and aims in life. The means–end chain model basically assumes that consumers will choose those products that are likely to produce the consequences that they consider positive and which fit their moral concept (Gutman, 1982; Reynolds and Gutman, 1988; Pieters *et al.*, 1995).

When applying the laddering technique, the interviewer starts with a question about the specific topic. For example, if the innovating company is interested in ways of improving its online banking services to make them attractive to current nonusers, the interviewer would ask nonusers what they consider relevant in online banking services and what they regard as the main advantages and disadvantages of online banking. Next, the interviewer would raise a series of "why" problems to identify the linkages between the nonusers' answers, their basic belief systems, and their motives for rejecting online banking services. If the respondents provide an answer to the first why question, the interviewer then generates a new "why" question based on the answer, moving step by step from the evident and immediate to the more hidden

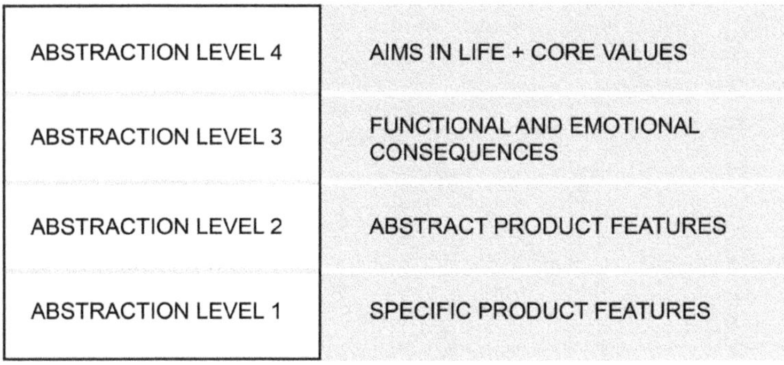

Figure 30: Categorization of latent needs.

and abstract reasons for the respondents' rejection of online banking services. To prevent the interview from becoming too interrogative or monotonous, the formulation of the questions can vary slightly (e.g., Why is feature X important to you? What does X mean for you? What do you associate with X?). Whatever the case, it is important that the interviewer is aware of the different levels of abstraction (see Figure 30) and aims at moving up the abstraction level through professional questioning. This method helps respondents think about their consumption practices in value-related terms. Finally, summarizing all of the respondents' identified reasons allows the innovating company to create a hierarchical value map that depicts the key reasons in their different levels of abstraction. The value map also shows the dominant connections between these abstractions (Jägel *et al.*, 2012; Cuny and Opaswongkarn, 2017).

Figure 31 illustrates the possible outcome of this procedure. It shows the product associations on different abstraction levels and the interrelations. For example, it could depict the link between online banking transactions' lack of an official receipt and several ladders of association networks leading to safety and control concerns. Based on such insights a provider of banking services could generate ideas on how to redesign their service offering to reduce some aspects of customers' safety and control concerns by providing evidence of payment (Kuisma *et al.*, 2007).

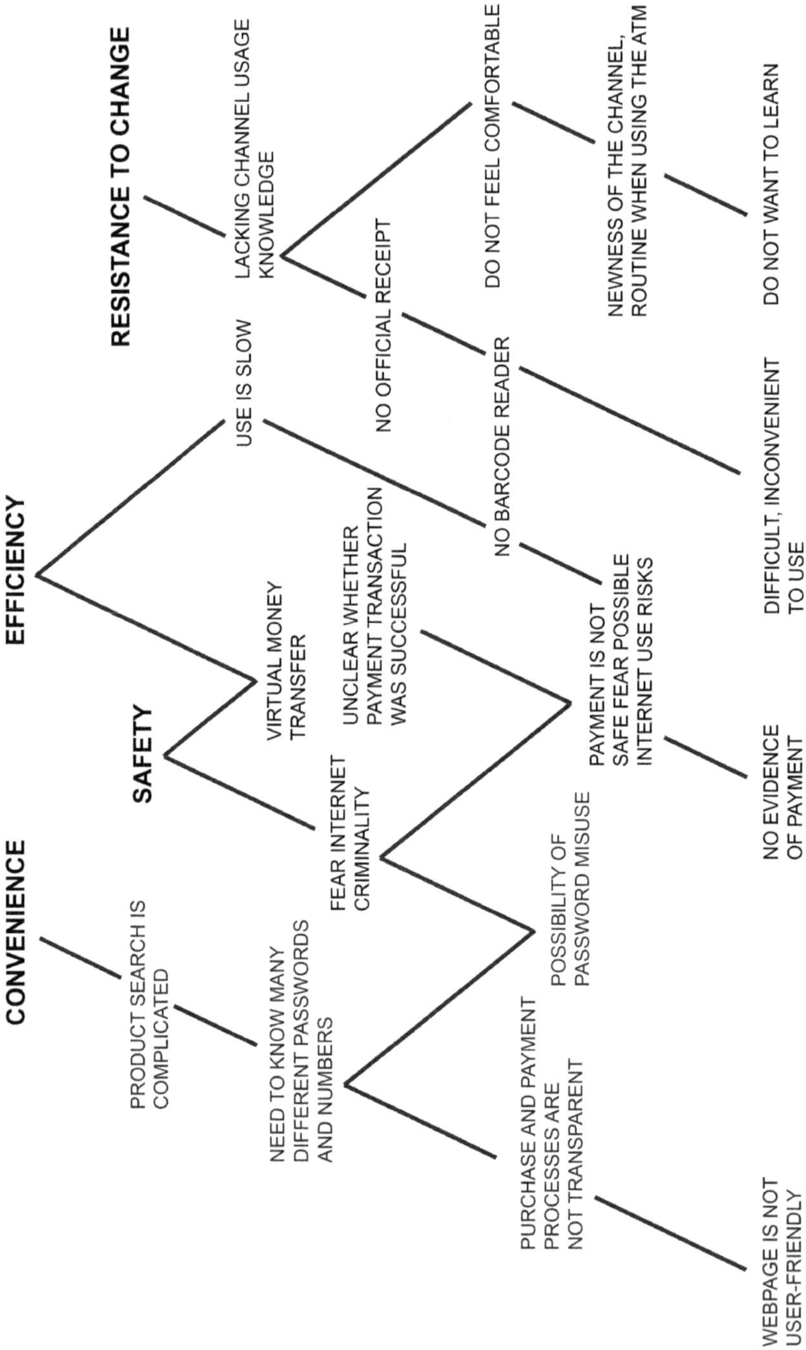

Figure 31: Barriers to online banking (Kuisma *et al.*, 2007).

Customer integration in action: Unveiling solution quality perceptions of B2B customers

The laddering technique is generally understood as a tool to connect products and consumption practices with a customer's self. This technique therefore seems to be potentially more valuable for B2C than B2B worlds. Contrary to this perspective, the following case shows that the laddering technique can also function as a powerful tool for understanding B2B customers.

Macdonald *et al.* (2016) investigate business customers' judgement of integrated product–service solutions with regard to medical devices, printing, pharmaceuticals, and building products. Some manufacturers offering such product–service systems appear to meet customers' needs very well and profit from these systems. However, on average, these manufacturers are less profitable than those with a pure product focus. This is at least true in the introductory phase, in which the latter seem to struggle with understanding the details that make solutions valuable for their customers. Macdonald *et al.* (2016) use the means–end chain approach to identify the link between solution quality, individual value-in-use perceptions, and organizational goals. This helps such manufacturers understand the relevance of product–service solutions' ability to contribute to individual goals, such as an employee's perceived stress reduction, and to collective ones, like faster problem solving.

Table 11 summarizes the key strengths and weaknesses of the technique.

Table 11: Strengths and challenges of the laddering technique.

Strengths	Weaknesses
• Largely unbiased consumer insights from passive lurking approaches • Plentiful and easy to obtain data that can be accessed unobtrusively • Data on user emotions toward products, as well as on user innovations and improvement suggestions	• Identifying appropriate online groups, as well as filtering and analyzing the relevant topics they discuss, is time and resource consuming, although some of the tasks can be outsourced to software programs • There are vibrant discussion groups for some product categories, not for others — netnographic approaches are thus not applicable to all the product categories

(Continued)

Table 11: (*Continued*)

Strengths	Weaknesses
• Qualitative research approach provides complexities and subtleties of specific customers, customer types, and the different facets according to which they perceive the specific topic	• False user identities can compromise insights • The small sample size allows no quantitative statistical inferences regarding the target group

3.8 Critical incident technique

The critical incident technique (CIT) is a sensitive, retrospective technique for gaining insight into customers' product experience through in-depth personal interviews. Its original intent was to identify failure incidents in consumers' prior service encounters. Its value in an innovation context lies in understanding the positive and negative incidents that change customers' behavior, or attitude, toward a service encounter. Understanding these incidents helps an innovating company improve services and create new services. While this method has its roots in the service field as a way to identify problems in service interactions, it can also be deployed to study customers' product experiences (Bitner *et al.*, 1990; Thomas and Bostrom, 2010).

A critical incident is any event that is sufficiently complete in itself to allow inferences about (1) what happened before, during, and after the incident; (2) the individuals involved; and (3) the intentions and consequences. A CIT has a tendency to often focus on events that consumers perceive as failures of their service or product offering usage (Flanagan, 1954; Chell, 1998).

Carrying out a CIT involves four sequential steps: setting the objectives, writing interview guidelines to gather critical incident data, gathering the data, and the data analysis and interpretation. The specifics of a CIT lie in the third step. In this step, data are gathered in a 2-h interview whose context is not predetermined but unfolds from the participants' perspective and reflects their experiences through their usual way of thinking about the incident. The method builds on the participants' capacity to narrate stories of incidents; it thus hinges on their eloquence and verbal skills. This makes the procedure challenging, as some participants might be unable, or unwilling, to take the time to tell a complete story when describing a critical incident. The interviewer starts by invitation the participants to think of a positive or negative experience with a service or product and then asks for details about the situation, the circumstances leading up to the situation, the consequences, and their feelings (e.g., Please think of a time when you had a particularly dissatisfying interaction with a supermarket: What happened? What was the situation like previously? What were the results of the situation? Why do you find the situation dissatisfying? When did it happen?). The interviewer encourages the participants to report the event in a natural way and share their perceptions while pondering the possible causes of the participants' actions and feelings. The interviewer keeps asking questions that help him or her gain an understanding of the incident and will clarify the details and possible inferences. In other words, the content unfolds from the respondents' perspective.

The data analysis and interpretation phase follows the basics of qualitative research techniques. The innovating company categorizes the incidents through carful and repeated readings and by sorting these incidents into groups (categories) of similar reported experiences and feelings. Thereafter the company labels these categories according to the nature of the similarities. Besides providing qualitative insights into the categories and the subcategories, content analysis can also provide quantitative information on the number of times the (sub)categories were mentioned and the lengths to which they were elaborated on. This information provides some indication of the (sub)categories' relevance for the participants (Bitner *et al.*, 1990; Gremler, 2004). Nevertheless, these quantities do not necessarily reflect the information's relevance for the whole customer base. Only quantitative deductive research designs would allow such conclusions. A CIT is qualitative by nature and is therefore an exploratory inductive method. For a more detailed discussion of the value of qualitative and quantitative designs, please see the info box in Chapter 4.

Research note: Interpreting critical incidents

Gremler (2004) reviewed 141 CIT studies and found that all but 11 use content analytic methods for their data analysis. He bemoans the lack of contextualization in the data analyses and explanations of reasons for experiences and feelings from the respondents' perspective. Analyses tend to regard respondents' stories as reports, striving to deduce facts about critical incidents through the researcher's interpretation of events, instead of understanding the respondents' account of the reasons for certain events happening. A combination of ethnographic and narrative approaches could reduce this problem and facilitate the collection and interpretation of emotions related to incidents.

CIT applications often focus on a single, rather short-term, incident interpreted in isolation, which aggravates the above shortcoming. Analyzing the same respondent's account of different incidents in a given context, thus capturing multiple instances of a specific incident type, might capture customer–firm relationships better for a better understanding. Variations of the CIT, such as the sequential incident technique (Stauss and Weinlich, 1997) and the switching-path analysis technique (Roos, 2002), address these deficits.

This approach's downside is its retrospective nature, which means it is prone to recall bias and memory lapses. The participants have to recall events, although their memories tend to fade over time. The participants therefore find it easier to discuss recent events and would rather report these, than those dating from further back. The respondents must be willing and able to describe an incident truthfully and in the required detail. Further, the innovating company has to decode and interpret the story correctly. This method focuses on retrieving information on critical incidents but does not provide insights into common, everyday practices and routines that shape customers' product and service experience. The specific focus channels views of extraordinary circumstances, which are not representative of normal encounters with a service, consumption, or use situation. The related events might be ones that the majority of users never encounter.

At the same time, the value of this approach lies in its scrutiny of rare events that a customer finds memorable and considers impactful and

Table 12: Strengths and challenges of the critical incident technique.

Strengths	Challenges
• The idiosyncratic participant perspective allows rich insights into participants' values and perspectives • Scrutinizes the causes and consequences of rare, but impactful, events • A rather inexpensive method • The flexible rules allow modification to meet a topic's requirements • The qualitative research approach allows for identifying the complexities and subtleties of specific customers, customer types, and the different facets within which they perceive the specific topic	• Sensitive to recall and reinterpretation bias • Everyday situations are ignored • Telling a complete story of an incident may be perceived as effortful and memory lapses might be replaced with fantasy • The small sample size allows no quantitative statistical inference regarding the target group • The data interpretation is a challenging task

decisive for future encounters with a specific service or product offering. Such events can thus have a critical effect on a system. Negative critical incidences highlight features or elements of a service encounter or product offering that might make the whole offering vulnerable (Table 12).

3.9 Zaltman metaphor elicitation technique (ZMET)

The Zaltman metaphor elicitation technique (ZMET) is a method used in personal in-depth interviews to deepen an innovating company's understanding of how customers experience constructs. These constructs are their opinions, feelings, and attitudes toward products, product categories, or service encounters. By combining personal interviews with elements of nonverbal expression (photos, pictures), the ZMET reveals the meanings and personal relevance that the participating customers associate with the selected constructs. This additional form of expression facilitates the articulation of complex meanings and subconscious feelings about products that are difficult to share verbally. The pictures and their symbolism also support a metaphorical reasoning about complex experiences and meanings.

The ZMET first asks the participants to take photos or to clip and collect pictures from magazines, indicating what a specific topic means to them and what they like or dislike about it. Roughly 2 weeks later, the participants explain their choice of photos and the feelings associated with the photos. The feelings about and opinions of the topic represent constructs, which the interviewer notes. Each photo is understood as a representation of one or more constructs that provide insights into the personal meanings customers attribute to a specific product. After the interview, the interviewer asks the participants to draw a map showing the connections between the important constructs. This collage represents a visual summary of the participants' thoughts on the topic and is the result of the evolution and maturation of their thinking about the topic throughout the 2-h interview session (Zaltman, 2003; Zaltman and Zaltman, 2008).

Christensen and Olson (2002) used the ZMET to identify the core experiences that consumers seek when mountain biking. These scholars identified four collective themes: riding for the thrill, connecting with a group, escaping, and nature. These exemplary results of a ZMET application show that it does not provide insights that can immediately be turned into new product ideas. Its value rather lies in exploring the consumers' unconscious mind beyond the surface-level thinking to understand why they think and act the way they do. This information provides the basis for innovation teams to put themselves in their customers' place to create products that are relevant from this perspective (Table 13).

Customer integration in action: From caterpillar to butterfly

An interior architecture firm used the ZMET to understand the needs of hospitals' end users before designing the Children's Hospital in Pittsburgh. Through the visual drawings and the end-users' stories, the firm learnt that transformation is a common theme for patients and their relatives, including transformation from being sick to healed, from anxiety to calmness after surgery, from the well-known area of one's home to one where the children are more isolated for a time. The architecture firm included transformation in the hospital's design elements, such as butterflies in the primary entrance and poems that give a positive feeling in the waiting areas, to load this transformation with positive feelings (Zaltman and Zaltman, 2008).

Table 13: Strengths and challenges of ZMET.

Strengths	Weaknesses
• Insights into customers' thought processes, motives, the personal meanings they attribute to products, and their complex interaction • Integrating pictures into the interviewing process expands the ways customers can express their opinions and feelings • The qualitative research approach allows for identifying the complexities and subtleties of specific customers, customer types, and the different facets in which they perceive the specific topic	• The outcome hinges on the participants' visual and metaphorical capabilities, as well as their verbal skills • It is time intensive for the participant (preparation and interview) • The interviewer needs experience with the ZMET technique to use it properly • The small sample size does not allow quantitative statistical inferences about the target group

3.10 Repertory grid

The repertory grid technique uses a distinctive semistructured interviewing procedure aimed at exploring a product's key attributes as identified by the respondents, rather than imposing the investigating company's or researcher's terminology. It stimulates interviewees to discuss complex themes in their own words. The method builds on Kelly's (1955) theory of personal construct systems, which suggests that individuals construct their realities by developing personal construct systems to interpret experiences and understand the world in which they live. They use this system to navigate current situations and anticipate future ones. These constructions are not permanent but change according to customers' experiences and their mental processing of situations and interactions with objects and persons that surround them. Metaphorically, constructs are the spectacles through which customers see the world.

In the repertory grid technique, these constructs represent identified and ranked product attributes, which are put in a matrix that allows a single-grid analysis, as well as a multigrid analysis of data. Repertory grids allow for capturing customers' insights into the individual meanings of products or brands (i.e., the elements in the repertory grid terminology) and identify the thought processes that reflect the use of mental maps to categorize and classify products and product features. The repertory grid approach usually involves a relatively small number of respondents (20–30) and comprises three consecutive steps (Lemke *et al.*, 2011; Baxter *et al.*, 2014):

1. semistructured interviewing
2. analyzing the individual grids
3. analyzing a series of grids.

The first step, semistructured personal interviewing, follows a strict sequence of questions as outlined in Figure 32. First, the interviewer asks the interviewees to reveal their awareness set of the elements in a category. The awareness set is the part of the total set of elements in a category that a specific individual knows. Individuals can only closely evaluate those elements in this awareness set (Kotler and Armstrong, 2009). If the innovating company is a manufacturer of building products, it would ask B2B customers (existing clients, as well as potentials) to name the brands that come to mind when they think of building products. Thereafter, the interviewer chooses three of the brands (=triad) and asks the respondents how these building product brands are similar to and different from the third product in terms of performance. If the respondents mention, for example, the quality, the interviewer gauges what they mean with the word "quality." In this way, the interviewer generates the constructs, which then represent customers' subjective perspective of the elements. The interviewer can also employ the laddering technique in conjunction with the above elicitation technique. Laddering allows the respondents to elaborate on the elicited constructs by building on general probing questions (i.e., how and why) to drill down into the meaning and interpretations of these constructs. Subsequently, the interviewer follows the same triadic elicitation process for the next triad. The interview is completed when the interviewer has gathered insights into and evaluations of 8–10 constructs or until the respondents are unable to provide further constructs (Jankowicz, 2004).

STEP 1 - FIND ELEMENTS

STEP 2 - CHOOSE TRIAD

STEP 3 - FIND CONSTRUCT

	A	B	C	D	E	F	
EASY 1							DIFFICULT 5

STEP 4 - EVALUATE THE TRIADS

	A	B	C	D	E	F	
EASY 1	5		4			1	DIFFICULT 5

STEP 5 - EVALUATE THE REMAINING ELEMENTS

	A	B	C	D	E	F	
EASY 1	5	2	4	5	2	1	DIFFICULT 5

STEP 6 - REPEAT STEPS 2 TO 5 WITH NEW TRIAD

	A	B	C	D	E	F	
EASY 1	5	2	4	5	2	1	DIFFICULT 5
RELIABLE 1	4	1	2	3	3	1	NOT RELIABLE 5

Figure 32: Interviewing steps for deducing elements and assessing constructs based on triads.

Repertory grids build on the assumption that personal constructs are bipolar by nature. After, the respondents have revealed the constructs, they are asked to name the contrast pole. Next, the respondents evaluate all the elements against the generated individual bipolar constructs, starting with the triad's elements. The interviewer subsequently asks the respondents to rate the elements from high to low performance on a numerical scale. For example, the respondents evaluate seven products' ease of use on a five-point rating scale, with 1 signifying "very easy to handle" and 5 "difficult to handle."

Instead of the triadic sort method, the innovating company can provide the constructs (and even the elements) if the research interest focuses on understanding a given set of constructs (elements) better. The advantage of providing constructs and elements is that there are higher commonalities between the different respondents provided elements with large differences decrease the problem that, in a triad, the combinations of elements can be quite similar and thus hamper the ability to elicit meaningful constructs. However, supplied constructs/elements detract from the charm of this method, which lies in taking the respondents' perspective (Tan and Hunter, 2002).

In the second step, the individual grids are analyzed. The received information is inserted in a grid matrix, with the rows and columns evolving during the interview. This matrix contains elements, constructs, and links. The elements are listed at the top of the matrix. The elements form the rows. The links refer to the elements and constructs and represent numbers indicating the extent to which a construct is present in an element. The respondents' ratings of each element link the elements to the constructs (Easterby-Smith, 1980; Tan and Hunter, 2002).

The individual grids allow for calculating the construct differentiation, the construct centrality, the construct preferences, and the construct variability to discover latent needs based on the constructs' and elements' individual relevance. The number of constructs in a repertory grid shows the cognitive differentiation, with many constructs depicting a highly differentiated grid. Cognitive differentiation is indicative of the respondents' involvement in the product category, with highly differentiated grids indicating high involvement. When their involvement is high, customers compare more features in more detail than when their involvement is low. In high involvement settings, innovative features distinguish a product from others, giving it a good chance of being preferred. Low involvement settings are more prone to making innovating difficult. Respondents showing high construct differentiation are therefore potential target groups for innovations based on latent needs (Dholakia, 1997; Zaichkowsky, 2013).

Construct centrality represents a construct's importance relative to all the other constructs. Central constructs are highly correlated with other constructs (Kelly, 1955). Table 14 presents a correlation matrix for manufacturers of metal-cutting machines. The correlations show "custom-made complete solutions" as the construct with the highest correlation with other variables and thus, the most central construct. High construct centrality is an indicator that the construct represents a latent need.

Construct preferences also provide information on each construct's perceived desirability, which is revealed by studying the row means. In Table 15, "cutting quality" has the best scores. The variance between each construct's elements shows whether the respondents rate the elements similarly (low variance) or differently (high variance). The higher the construct's variability, the more important it is as a latent needs focus area. In Table 15, "low downtime" has the highest variability.

The third step involves analyzing a series of grids. The individual results are captured in one matrix to examine the similarities and the differences across the respondents. To reduce the number of constructs in the grid, the analyst identifies the themes or categories underlying the constructs, thereafter classifies the common constructs into categories, while registering their relative frequencies. Subsequently, the analyst identifies the key constructs, which are presumed to indicate latent needs, and builds a good focus area for innovation activity. These key constructs are high-frequency constructs (i.e., constructs mentioned by at least 25% of the respondents), as well as constructs showing higher than average variability. To assess the average variability, the analyst calculates the average normalized variability of a construct. The normalized value prevents too much relative importance being attached to those constructs mentioned by a low total number of respondents. If a construct's average normalized

Table 14: Correlation matrix for manufacturers of metal-cutting machines (fictive data).

No.	Constructs	Construct numbers 1	2	3	4	5	6	7	8	9	10
1	Serious and reliable	—	—	—	—	—	—	—	—	—	—
2	NPD capabilities	**0.84**	—	—	—	—	—	—	—	—	—
3	Active relationship management	**0.93**	0.57	—	—	—	—	—	—	—	—
4	Easy-to-use machines	25	0.51	0.31	—	—	—	—	—	—	—
5	Intuitive interfaces	0.36	0.62	0.22	**0.88**	—	—	—	—	—	—
6	Low downtime	0.66	0.22	0.52	0.41	0.33	—	—	—	—	—
7	Attractive price level	0.53	0.34	0.44	0.17	0.24	0.29	—	—	—	—
8	Custom-made complete solutions	0.78	**0.94**	**0.91**	**0.83**	0.16	0.55	–0.11	—	—	—
9	Proximity to customer	0.66	0.51	**0.95**	0.07	0.58	**0.94**	0.03	**0.82**	—	—
10	Short response time	0.62	0.62	**0.88**	–0.37	0.20	**0.96**	0.24	0.53	**0.95**	—
11	Cutting quality	0.69	0.55	0.60	0.51	0.08	0.21	0.18	**0.95**	0.46	0.16

Table 15: Construct preferences for manufacturers of metal-cutting machines (fictive data).

No.	Constructs	Min.	Max.	Mean	SD	Variance	Variance (%)
1	Serious and reliable	1	5	3.17	1.72	2.9584	10.04
2	NPD capabilities	1	4	2.67	1.21	1.47	4.98
3	Active relationship management	1	5	2.67	1.51	2.27	7.69
4	Easy-to-use machines	1	5	3.00	1.79	3.20	10.86
5	Intuitive interfaces	2	4	3.17	0.75	0.57	1.92
6	Low downtime	1	5	2.83	1.83	3.37	11.42
7	Attractive price level	1	5	3.67	1.51	2.27	7.69
8	Custom-made complete solutions	3	5	4.17	0.98	0.97	3.28
9	Proximity to customer	1	5	3.50	1.52	2.30	7.80
10	Short response time	2	3	2.83	1.17	1.37	4.64
11	Cutting quality	1	3	2.50	0.84	0.70	2.38
				34.17	**14.82**	**21.43**	**100**
				2.85	**1.24**	**1.79**	—

Table 16: Strengths and challenges of the repertory grid method.

Strengths	Challenges
• Encourages respondents' reflection on the personal meaning of and preferences for products and product attributes (if constructs are generated by respondent) • Profound insights into individual thinking	• Relevance of constructs for product choice is not explicit (notably if constructs are provided by researcher) • Respondents might find it difficult to produce enough elements and constructs for the triads method to work properly

(Continued)

Table 16: (*Continued*)

Strengths	Challenges
• Helps identify product characteristics that respondents consider important and that vary from competitor to competitor • The qualitative research approach allows for identifying the complexities and subtleties of differences between specific customers, customer types, and the different facets with which they perceive the specific topic	• The artificial way of questioning is strenuous for the interviewee and the interviewer • A relatively time-consuming personal interviewing technique • The results are dependent on the participants' verbal ability • The small sample size allows no quantitative statistical inference to be made regarding the target group

variability is above 100 divided by the average number of constructs across all the respondents, the construct is important for respondents and should be classified as a key construct (Goffin *et al.*, 2010) (Table 16).

Chapter 4

Approaches to close the solution space: Customers as evaluators

4.1 Preference tests and test errors

Customer integration approaches to close the solution space strive to reduce uncertainties regarding a new product's customer value by evaluating its adoption likelihood. These approaches include concept, product, and market tests. The different test forms correspond to different stages of the NPD phase. In the early stages, an innovating company uses tests to differentiate strong concepts from poor ones before investing substantially in the former's further development. Throughout the development and pre launch, this company uses product tests to optimize the product's customer value. Prior to its launch, market tests measure the entire marketing mix's appropriateness for the product. Basically, all three tests strive to select the most promising solutions to carry forward to the marketplace and to estimate the sales.

The specific NPD/NSD stage in which a test is carried out influences the results. Figure 33 shows the trade-off between prognostic validity and the product design's scope. In the early stages of NPD/NSD, the final product's concept is still fuzzy, which reduces the prognostic validity of the test results. At the same time, the product design's scope is still wide

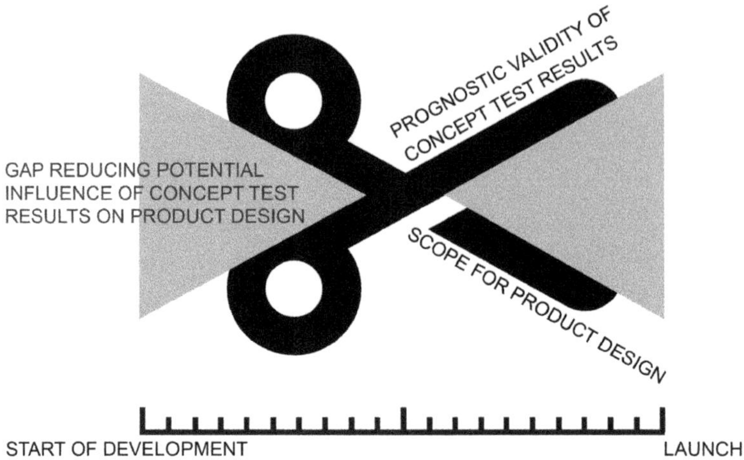

Figure 33: Estimating new product market performance based on test results. (Adapted with permission from Hagstotz and Schmitt-Hagstotz (2008).)

open. Throughout the entire NPD/NSD process, clear indications of the right product design have the highest leverage to influence the final product design. Throughout the following NPD stages, NPD/NSD teams gradually decrease the product design scope, test the technical feasibility, and adapt the concept accordingly until they eventually freeze the design. While these activities reduce the product design scope, they make the final product's content and characteristics clearer. With the product moving from a rough concept to a functional prototype and beyond, the possibilities for accurate testing increase. The closer the product is to its final form, and thus ready for launch, the higher the accuracy of the tests that measure its market attractiveness and the less such tests can effectuate design changes. If a need for profound design changes occurs during the late NPD/NSD stages, these changes can be quite costly and time consuming. The NPD/NSD process might need to be reiterated, which means the time to launch could be postponed.

A meaningful approach to cope with the trade-off between the ability to influence the product design scope and the prognostic validity of the concept test results is to carry out several tests throughout the innovation process. Start with stripped concept tests and later progress toward more testing, more elaborate concept versions that help refine the concept step by step and enable increasingly finer-grained feedback on customer preferences and the new product's market attractiveness.

As illustrated in Figure 34, tests intended to close the solution space are prone to two error types. The first error type applies when a concept yields high adoption intentions in concept tests, but the final product fails to reach this acceptance in the market place, indicating that a no-go decision would have been more appropriate (=false positives). These errors are unpleasant, because the company wastes valuable resources on developing products with a low potential. Since these concepts are likely to finally make it to the market, this type of error is easy to calculate. The second error type, false negatives, is better concealed, because the concept rates poorly in a concept test, although it would have developed into a successful new product. These missed opportunities lead to losses of income and can result in competitors that do realize these concepts pre-empting the company. It is difficult to gauge the number of concepts that are abandoned or fall into the second error type category. Nevertheless, researchers and practitioners regard this error type as a great challenge, especially for radical innovation projects (Klink and Athaide, 2006; Gassmann and Schweitzer, 2014).

EVALUATION IN CONCEPT TEST \ SUCCESS IN MARKETPLACE IF REALIZED	LOW	HIGH
LOW	CORRECT NEGATIVES	FALSE NEGATIVES
HIGH	FALSE POSITIVES	CORRECT POSITIVES

Figure 34: Errors in NPD/NSD-focused preference tests.

Research note: The danger of false negatives in concept tests for radical innovations

Communicating a radical innovation idea to the customers appropriately at an early stage is challenging, partially because the idea is still rather vague and the innovating company itself is still figuring out its potential and specifics (Beckley *et al.*, 2012) and partially due to the customers' imagination difficulties.

In concept tests, the product information is usually incomplete and consumers cannot test the product to experience its use. They have to roughly gauge what the final product might look like, feel, and how it will function. Owing to the nature of radically new products (RNPs), customers have no, or very limited, use experience with similar products. Consequently, they cannot use preferences based on prior experience but must construct preferences on the spot when evaluating RNPs. Hoeffler (2003) showed that customers are more uncertain about the benefits, drawbacks, and social implications when evaluating radical concepts rather than incremental ones, as well as about their ability to evaluate them correctly.

Schoormans and colleagues (1995) add that preferences for radical innovations are unstable over time, with customers evaluating options differently in consecutive evaluation tasks. In a similar vein, Bohlmann *et al.* (2013) highlight the dynamic interplay of customer needs and evolving innovations that turns customers' product preferences and expectations into a moving target, changing with the changes in market offerings.

Veryzer (1998) provides illustrative cases in which customers were unable to understand the benefits of radical innovations and thus contributed meaningless information in concept tests. Urban *et al.* (1997), Zhao *et al.* (2012), and MacDonald (1995) stress customers' evaluation difficulties due to their lack of imagination regarding visualizing a potential future usage scenario for radical innovations. This lack of imagination results in less favorable product evaluations for radical than for incremental innovations. Taken together, this research alludes to the difficulties of retrieving accurate customer preference statements regarding RNPs and shows that customers' tendency to ask for the familiar and reject radical innovations in preference tests is likely to provoke a false negative error in respect of radical innovations.

Generally speaking, concept tests are less likely to provide accurate results than product and market tests, as they are carried out at the early stages during which the way the final product feels and performs is still fuzzy. In addition, at that time, the current market situation does not necessarily reflect the market situation at the time the concept is developed into a product, while customer needs might have changed in the interim. Moreover, radical concepts are more prone to errors in concept testing than incremental ones, as customers per definition lack use experience for such products and find it more difficult to infer missing information and to imagine radical product concepts' potential future use. Yet, a sound choice of respondents and the appropriate use of stimuli can help reduce the likelihood of errors in concept, product, and market tests, as well as provide realistic insights into customer preferences.

4.2 Qualitative and quantitative preference tests

The innovating firm can use qualitative or quantitative methods to evaluate customer responses to concepts or products. While both methods aim to help facilitate decision making in NPD/NSD, their contribution differs in so far as the qualitative methods focus on diagnostics and the quantitative ones on description.

Qualitative concept/product tests are rich in insights into reasons for customers liking or disliking a concept/product, rather than merely providing numbers of likes and dislikes. As in any qualitative research, qualitative concept/product tests are rooted in the humanistic ideal of meaningful understanding. The researcher investigates a small number of cases thoroughly in an open, nonstandardized, intersubjective way and without predetermined assumptions. Techniques to relate to the participants are adapted to the circumstances and individual cases. The meaningful application of these tests requires well-trained and experiences interviewers who can gather insights beyond evident first answers. The outcome of such tests allows no clear course of action recommendation (go/no-go) but rather needs discussions and interpretations in the NPD/NSD team for a common interpretation of the data to develop.

The tests provide the innovating firm with information that helps it explore behaviors and attitudes toward the new concept and find ways to improve or change the concepts/products. Such tests also provide a first indication of customers' interest in the innovation. In B2C companies, qualitative concept/product tests usually take the form of focus groups. In B2B companies, individual tests are more common. The sales representatives sometimes carry out such tests when vising their clients. The evident advantage of this approach is that sales representatives have relatively easy access to key informants in the client organization and that it is relatively fast and cost effective to send them to clients. The substantial downsides are, however, that (1) sales representatives usually lack market research expertise required to carry out these tests appropriately, (2) clients might provide opportunistic answers as sales reps are not neutral dialogue partners, and (3) sales reps are generally paid to generate margins and not to carry out interviews, which makes them reluctant to carry out such tests.

Quantitative concept/product tests capture customer/product feedback from a representative, large customer sample. The aim is to test the adoption intention in various, potentially attractive, market segments by means of statistical methods based on large data sets. The quantitative approach strives to exclude decision-makers' subjectivity.

A quantitative product/concept test allows for determining the frequency and distribution of buying intentions and to relate these to target group characteristics. It is a deductive approach to verify, or falsify, hypotheses regarding the most attractive target group, the features this group values most, and their willingness to pay for a certain quality level of these features. Based on quantitative data, clear information is provided on the attractiveness of the concept/product for key audiences, which facilitates the further choice of actions to optimize the concept/product and target potential customers.

Quantitative tests are highly structured and have a predominantly closed-end question format. The key interest in quantitative tests is to obtain measurable respondent data that allow for drawing conclusions regarding the target population, such as the percentage of people interested in adopting a new product in the potential key target group. Quantitative tests allow for quantifying the adoption intention and for identifying its most important drivers. However, it does not allow the innovating company to discover possible factors that could explain the interaction between the drivers and the adoption intention of which the company is unaware. Since the test simply measures the importance of items that the innovating company has included in a questionnaire, items that the company has not thought of are not part of the test.

The formulation of the questions and the arrangement of the layout need thorough consideration, and a pretest should be carried out to check for clear and concise formulation. Contrary to qualitative test situations, the format and questions cannot be changed during the test and the person executing the test with the participants needs very little training and experience. Data analysis builds on statistics and is thus less prone to objectivity problems.

Infobox 4.2. Customer integration in action: Best of both worlds through combination

A manufacturer of mobile phones and apps for the elderly carried out a qualitative concept test with five focus groups to study their perceptions of two new applications: an interactive calendar and a photo-sharing tool. Generally speaking, the company's mobiles attract elderly consumers who are not comfortable with technology and often ask their children or grandchildren for help with keying in new contacts, changing certain settings, and similar tasks. The interactive calendar application was planned to allow relatives to fulfill some of these tasks remotely via their own mobiles or other devices, while the photo-sharing tool allowed them to share photos while talking about them on the phone. These applications involve two different user groups: the elderly and their relatives. To gather insights into both groups, two focus group workshops only included elderly consumers, two only included relatives, and one included families of elderly and their relatives. These focus groups captured not only input on the design and functioning of the concepts but also rich insights into motivations for using such applications and into the application situations. The focus groups also provided many ideas for improving the applications' perceived customer value. These insights led to intensive revisions of the

(Continued)

(Continued)

concepts, to new concepts tested qualitatively, and a prototype smartphone subsequently tested by means of a large sample. The qualitative tests helped change the concept several times to ultimately arrive at a promising solution. The quantitative tests provided insights into the value the different market segments and user groups attribute to the prototype. Based on the quantitative tests, the mobile phone manufacturer identified the most attractive segment, the optional features this segment valued most, and the positioning strategy. The endeavor resulted in a highly successful product introduction.

Figure 35 shows typical sample questions and depicts the central pros and cons of qualitative and quantitative concept/product tests.

From our discussions with managers in B2B companies, we learnt that they far too often find a qualitative test sufficient to grasp a new offering's market potential. We strongly caution against using only one test for concept/product optimization and market estimation: Qualitative tests are unable to deliver a representative basis for market estimations, while quantitative ones have deficiencies regarding delivering the required depth of understanding of potential customers' way of thinking about certain concepts and these concepts' potential to meet or not to meet these customers' expectations. There are also arguments against quantitative testing in B2B markets, such as the small target groups, the high costs of samples (e.g., machinery used in in-house tests). Clever test design, such

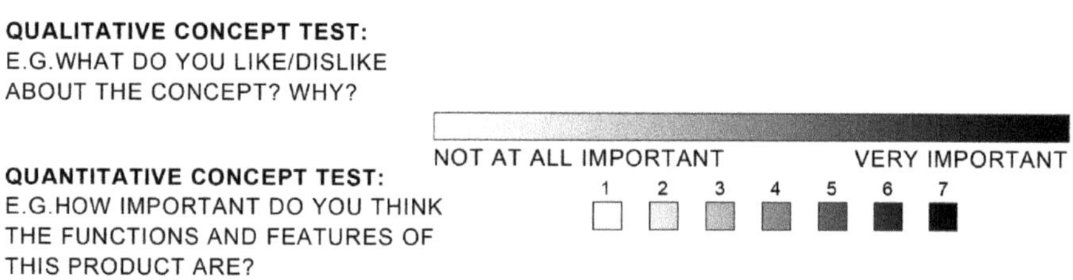

Figure 35: Sample questions of qualitative versus quantitative concept/product tests.

as studio tests for expensive machinery may allow testing acceptance with more prospects in a cost-efficient way.

We therefore suggest a three-step procedure:

1. screening concept tests
2. qualitative concept optimization tests
3. quantitative concept tests.

Screening tests aim at identifying interesting concepts from a plethora of presented ideas. Screening tests allow for making a rough estimate of the possible customer reactions to a new product. These tests take place at a very early stage, when the concepts are no more than short descriptions of the key benefit and allow respondents to screen 30–40 of them in one go. The concepts are often limited to a few words (for new convenience food, this would be, e.g., teaching children about food, to have fun with food, to value the ingredients in food, and to enjoy certain food) to allow the screening of such a large number of concepts. The evaluation task thus focuses on a single, quantitative question from which the adoption intention can be inferred (usually a liking question as the concepts are as yet too imprecise to ask purchase intention questions, e.g., "How interested are you in the following product concept on a scale from 1 (=not at all interested) to 7 (=very interested)?). The main goal of such screening tests is a first market-based filtering of the concepts to select the ideas with the most positive ratings and sift out low-rated ones, but an adaption of the test also allows for developing the ideas further. Examples are product ideas that the majority of customer raters rated low, but a few extreme raters rated very positively. Discussions could be held with these raters to understand their background and what they like about the product. These insights could help detect latent needs that might be attractive to a larger market segment (Durgee *et al.*, 1998).

In offline tests, 300–500 respondents are invited to such tests, while online tests' screenings usually occur in the form of a community voting for ideas generated through online idea competitions. These online tests provide fast feedback, decrease the cost of testing, and allow for testing more ideas with a larger number of respondents. Their merits also have downsides. Online idea voting has to be handled with caution, as it is often unclear whether the respondents represent the target group adequately and reveal their real preferences (e.g., some idea providers can activate friends to vote for their ideas in order to win a competition, voters can also hijack a competition and use it as a way to let off steam, or make fun of the company (e.g., by voting for ideas that they consider funny instead of for ideas that they consider attractive buying options).

After the first screening, qualitative concept optimization tests help identify the concept's uniqueness, make the concept statement clear and forceful, provide an understanding of the advantages and disadvantages, and help with the adoption and refinement of the concept according to these insights.

As soon as the key benefits of the concept become clear, as well as the way in which the concept delivers them, quantitative concept tests become an interesting option. In these tests, a minimum of 100 customers of the potential target market need to participate to allow valid statistical tests. A higher number is recommended to identify the differences in the perceptions of the target market's different subgroups. In B2C settings, reaching such numbers is pretty straightforward. In some B2B settings, where the actual target market population is smaller, the ability to carry out statistical analyses may be limited to a few nonparametric tests.

To sum up, Table 17 compares the key strengths and weaknesses of qualitative and quantitative preference tests.

Table 17: Strengths and challenges of qualitative and quantitative preference tests.

Quantitative test	Qualitative test
Strengths: • Large number of representative cases • Calculated adoption intensions provide statistical information for go/no-go decisions • Allows statistical calculations of market estimations based on test results	Strengths: • Allows product/concept optimization based on a rich understanding of preferences • Allows discovery of adoption-relevant factors of which a company is not yet aware • Rather inexpensive and time efficient

(Continued)

Table 17: (*Continued*)

Quantitative test	Qualitative test
Weaknesses: • Lack of exploration of concept optimization possibilities and adoption-relevant factors that the innovating company did not think of in advance • Not applicable if the customer base for a specific offering is small and the offerings are highly individualized, as in some B2B areas	Weaknesses: • Small number of nonrepresentative cases • Prone to subjectivity bias during data gathering and interpretation • Statistical calculations of market estimations not possible from test results

4.3 Respondent selection

Respondents represent a sample of the target population. The target population includes all individuals with relevant information for the innovating company. In concept tests, the target population is typically the new offering's target group. It is rather easy to describe who these prospects are in respect of incrementally new products, but in respect of radically new products, this task is more challenging.

Research note: Respondent selection for radical innovation projects

In radical innovation projects, the selection of the right respondents is a trickier task than in projects that focus on incrementally new offerings. The right target group for radical innovations and the best application of a new technology is usually not clear in the beginning. In incremental innovation projects, a respondent's prior experience with the product category increases the test result accuracy. In radical innovation projects, customers who are familiar with existing products, for example, customers with technical knowledge of petrol-based cars might find it more difficult to understand and trust a new technology, such as hydrogen fuel cell cars. These customers might base their quality judgements on criteria that

(*Continued*)

(Continued)

worked well for a familiar product category (e.g., the noise of a starting car as an indication of its engine's power), but that do not meet the criteria for a radically new product. Respondents with more experience with familiar products will thus not only have to learn new practices but also need to unlearn their habitualized old ones before they can correctly evaluate a radically new product (Urban *et al.*, 1992). In other words, if the similarities between existing products' and the radically new product's functions could be misleading, it is better to invite respondents with low levels of expertise of such existing products or product categories.

Many companies include respondents with a high level of innovativeness (i.e., a high propensity to adopt new products) in concept tests for radically new products, because they find it easier than the average customer to identify meaningful evaluative criteria for these products and to gauge their potential. This respondent selection may lead to overly positive concept/product evaluations, resulting in a proinnovation bias. Although these respondents are likely to be the first adopters of the new product, later adopters might be interested in other features. Focusing only on early adopters may result in offerings that neglect features that the main part of the potential future target market finds relevant (Moore, 1999; Peng and Finn, 2010). A better approach could be to take the respondents' different levels of innovativeness into consideration and run separate analyses for these respondent groups in order to understand their specific needs and requirements better.

Producing early prototypes that allow potential customers' R&D departments to experiment with them on their own is a good way of working with B2B customers on radical innovation projects. O'Connor and Rice (2013) find that customers then apply the innovation in areas which the company was unaware of, enabling it to identify promising markets and application fields.

Once the innovating company has specified the target population, it has to identify a sampling frame. A sampling frame is a list, or set, of directions for identifying individuals who represent the target population. Customer databases, telephone books, or lists from specialized address agencies may help compile a sampling frame. Depending on access to accurate lists of the target population, a smaller or bigger difference occurs between the target population and the sampling firm (sampling frame error). For example, sampling frame errors occur in telephone books that capture households in an area, because some households choose not to be listed, others are not yet listed because they have recently moved to the area, and still others are listed but have recently moved out of the area. The innovating company has to be aware of potential sampling frame errors and think of options to reduce these.

From the list of individuals in the sampling frame, the innovating company has to choose a sample as the respondents for the concept test, which it does by applying nonprobability or probability sampling techniques (Figure 36). Either probability sampling, or the nonprobability-based quota sampling technique, is a viable option for quantitative concept tests. The company is well advised to use one of several nonprobability sampling techniques for qualitative concept tests.

With probability sampling, each individual basically has a statistically determinable probability of being included in the sample, with this probability being different from zero. Further, each element of the sample frame has to be clearly identifiable; probability sampling therefore allows the statistical assessment of a sampling error. In simple random sampling, each individual is selected independently of other individual selections and has an equal selection probability. A typical example of this sampling procedure is for the innovating company to write all the names of the individuals of a given sampling frame on individual pieces of paper, to put these in a big box, mix them, and then select a certain number in a lottery-style drawing. A software-based alternative involves putting a number next to every name, then using

SAMPLING	
PROBABILITY SAMPLING	**NONPROBABILITY SAMPLING**
SIMPLE RANDOM SAMPLING	QUOTA SAMPLING
SYSTEMATIC SAMPLING	JUDGMENTAL SAMPLING
STRATIFIED SAMPLING	SNOWBALL SAMPLING
CLUSTER SAMPLING	THEORETICAL SAMPLING
	CONVENIENCE SAMPLING

Figure 36: Probability and nonprobability sampling techniques.

Mathematica, or other software with random number generation possibilities, and selecting the names behind the generated numbers (e.g., 2, 17, 83, 452). Simple random sampling in geographically widely spread target populations can be relatively expensive and time consuming, yet delivers valuable, representative samples.

Systematic sampling is another probability sampling technique in which the innovating company has a numbered list of the sampling frame. It first chooses a starting point randomly, e.g., number 73. If a B2B company has 240,000 potential clients for a new product in the sampling frame and wants to invite 600 to a concept test, it then invites every 400th starting from 73 (in other words, the companies with the numbers 73, 473, 873, 1,273, 1,673, 2,073, 2,473,...). If the numbers next to the company relate to certain characteristics (e.g., the company turnover in the last year, the number of years the innovating company has had a relationship with it, the company size), systematic sampling provides a better distributed sample relative to these characteristics (e.g., small and large companies if listed by size) than simple random sampling.

Stratified sampling requires dividing the sampling frame into two to six strata. Individuals within a stratum need to be homogenous regarding the characteristic chosen for the strata construction and heterogeneous regarding the individuals in the other strata. For example, a company could build three strata for its existing customers based on their turnover (clients with a turnover < €50,000, clients with a turnover between €50,000 and €100,000, clients with a turnover above €100,000). Another option would be to build two strata in order to differentiate between a company's existing customers and not-yet customers. The company then carries out simple random sampling within each group.

Cluster sampling requires the innovating company to find clusters that contain a variety of all the possible elements in a target population. If individuals within a cluster are heterogeneous and this heterogeneity is representative of the target population and covers the population's diversity well, the company can focus on a couple of clusters instead of sampling from all the clusters, while increasing the sampling feasibility and decreasing the sampling expense. The difficulty of this approach inevitably lies in finding accurate clusters from which the target population is drawn.

Contrary to probability sampling, the selection of individuals in nonprobability sampling is not carried out randomly but by the innovating company's more or less conscious selection. Except for quota sampling, nonprobability sampling does not allow generalization of the results in terms of the target population and is therefore not applicable for quantitative concept testing.

Quota sampling builds on control categories, or quotas, of elements whose percentage is known in the target population. For example, if the innovating company knows that the target population for a new product consists of 39% women and 61% men, the quota-based sampling of 100 respondents would mean choosing 39 women and 61 men. A problem with this approach is that the innovating firm might not have access to, or might even miss, relevant quotas, resulting in selection bias. Nevertheless, sophisticated quota sampling approaches allow for data quality that is close to probability sampling regarding the target population's representativeness.

Judgmental sampling is a nonprobability sampling technique in which the innovating company chooses certain respondents based on their fulfillment of certain criteria. For example, if a B2B company wishes to involve its most innovative clients in a qualitative concept test, the company first defines what it understands by innovative (e.g., clients who won an innovation prize are innovation leaders in their industry, who challenge R&D with their customer requirements). Thereafter, the company selects clients that — according to its judgement — possess one or more of the required criteria. This technique is quick, inexpensive, and convenient.

Snowball sampling is a good sampling approach if members of a target population are relatively rare, membership of a community is confidential, the members are difficult to identify, and they have well-functioning internal networks. As soon as the innovating company finds initial respondents from this target group, it asks them to identify others. After interviewing the others, the innovating company again asks for referrals. With one respondent referring another, this sampling technique results in a snowball effect that gives the technique its name.

Theoretical sampling (Glaser and Strauss, 1967) is a sampling approach for qualitative preference tests carried out in the form of individual interviews. Applying this approach, the innovating company first chooses someone who represents a potentially attractive market segment very well and then interviews someone who is very similar to this person. Next, it asks someone from a very different, but also potentially attractive, market segment, again someone similar, and again someone who is very different. In this way, the innovating company continues to gather feedback on the concept until it has a clear view of the best market segments and the main ways of improving the existing concept. When the company no longer receives any new information from interviews, because all the issues raised, or all the insights into the market segment values related to the concept have already been mentioned, this means it has conducted the right number of interviews and reached theoretical saturation. This procedure maximizes the differences and enables great coverage of the variation between different aspects. The innovating company thus receives an elaborated view of the concept's various facets and what the potential target markets think about it.

Convenience sampling's primary criterion for selecting respondents is obtaining a sample that the researcher can easily access. This sampling approach has a very strong selection bias. The respondents may be quite different from the target population and should therefore be avoided.

4.4 Typical question set

In product and concept tests, the participants are usually asked to evaluate concepts on a number of different dimensions (see Figure 37) encompassing their purchase intention, how much they like the concepts, and how much they believe in its superiority and problem-solving ability. Questions also include background information about the respondent. The respondents' purchase intention regarding nondurables, together with their buying frequency, is the primary criterion on which the innovating company bases its decision to either take a concept or product further toward launch or to kill it. Typical questions to measure purchase intention include: "How great is your willingness to purchase the product?" and "How often/ In what quantity would you buy this product?". Such questions are often complemented with questions on the respondents' willingness to pay ("What price would you be willing to pay for this product?").

A further, very important, set of questions is based on liking or disliking the product. These questions aim at gaining a feel for the product's

1.PURCHASE INTENT / BUYING FREQUENCY
HOW GREAT IS THE WILLINGNESS TO PURCHASE THE PRODUCT?

2.LIKING / DISLIKING
WHAT DO CUSTOMERS LIKE AND DISLIKE ABOUT THE CONCEPT?

3.COMPREHENSION / BELIEVABILITY
IS THE CONCEPT CLEAR AND BELIEVABLE?

4.PROBLEM SOLVING ABILITY
DOES THE PRODUCT SOLVE A PROBLEM OR ADDRESS A CERTAIN NEED?

5.IMPORTANCE
IS THE PRODUCT IMPORTANT FOR CUSTOMERS?

6.UNIQUENESS / COMPARISON WITH CURRENT OFFERING
DOES THE PRODUCT HAVE ANY ADVANTAGES COMPARED TO OTHER PRODUCTS?

7.BUYER CHARACTERISTICS
SOCIO-DEMOGRAPHICS AND PSYCHOGRAPHICS OF RESPONDENTS

Figure 37: Typical dimensions of test questions.

market acceptance. The overall extent of liking the product can be a key criterion for the go/kill decisions regarding very rough concepts for which adoption intention questions are not yet appropriate (see screening concept tests). In addition, specific inquiries into the reasons for liking or disliking help gather diagnostic insights into customers' reasons for judging a concept the way they do, and provide scope for improving a concept. These insights are prompted with questions such as: "What do you like and dislike about the concept?," "Why do you dislike feature XY?," and "How could the concept be improved to make it more interesting for you?".

Concept tests also include questions on comprehension and believability. The innovating company needs to know if the concept is easy to understand and if the respondents understand it correctly, because a lack of comprehension inevitably leads to wrong preference statements. Believability hints at the concept's communicability and its benefits. Respondents' feedback on comprehension and believability helps make products more convincing and understandable. Comprehension and believability are captured through objective comprehension questions and subjective feedback on questions such as: "How credible are the mentioned benefits of this product? What is more/less plausible? Why?".

Questions about a product's importance, problem-solving ability, and uniqueness provide insights into the new product's relevance from the potential customer's point of view. Important concepts relate to concepts that provide features and functions catering to customers' unfulfilled needs. A concept has to solve an important problem for the customer to be attractive. Further, current market offerings should not already solve this problem well enough, otherwise it lacks uniqueness. The following questions help gather information on the product's relevance for the customer: "Which need does the product address? How important is this need for you? How sure are you that the product will be able to satisfy this need? Are there other current market offerings that satisfy this need well? To what extent does this product differ from these alternatives?".

In addition to all these product-related questions, concept tests also include questions on buyer characteristics, such as the respondents' socio-demographics (e.g., gender, age, income) and psychographics (e.g., lifestyle, consumption intensity of similar products). Buyer characteristics indicate the concept's attractiveness for different market segments. This information allows for calculation of the market size and scenarios regarding variations in the market size if the concept is adapted one way or the other (Moore, 1982; Kotler and Keller, 2006; Peng and Finn, 2008, 2010).

Customer integration in action: Hilti's concept-testing approach helps introduce power tools successfully

Hilti regularly carries out concept tests in NPD projects. Applying a similar set of questions across various projects enables Hilti to compare the evaluative feedback on a current project with the concept test feedback on prior successful projects, as well as on projects with little market acceptance. These comparisons facilitate new product go/stop decision-making. In 2011, Hilti tested the AQA prototype, an impact wrench to set different types of stud anchors. The prototype included a sensor and algorithm measuring the angle changes in the tool's head, investigating the

(*Continued*)

(Continued)

correct initial stress, and automatically stopping the machine as soon as the correct level was reached. In addition, visual feedback informed the user whether the stud-setting process was correct (Figure 38).

The key questions for go/no go decisions are asked similarly throughout different Hilti concept tests to allow a comparison and questions on the first spontaneous impression ("You have tested the prototype regarding its setting behavior. What is your first spontaneous impression? Please state your answer by using a scale from 1 (not appealing at all) to 10 (very appealing)"). This is followed by the evaluation of the tested prototype (regarding criteria such as the overall setting time, ease of use, innovativeness, or self-explanatory usage — again indicated on a 1 (very poor) to 10 (very good) scale). Next the adoption intention is queried ("If the price was not an issue, how likely would you be to buy the tested prototype on a scale from 1 (very unlikely) to 5 (very likely)?").

Hilti also gathered general feedback on the interviewees' liking/disliking of the AQA prototype, along with their reasons for liking/disliking it. Another important part of Hilti's AQU prototype test was sociodemographic information of the interviewees' company (country, number of employees,

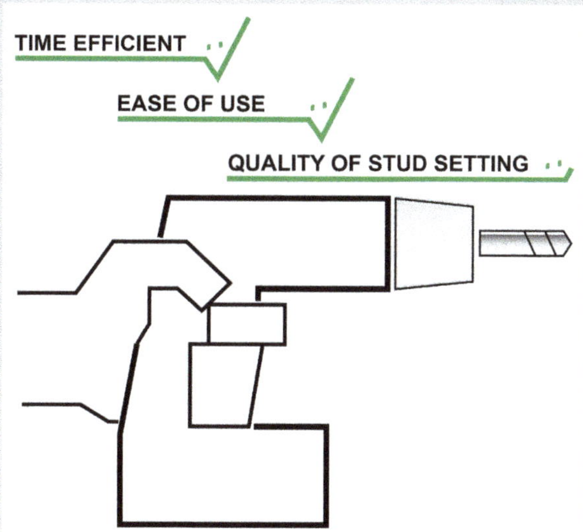

Figure 38: Concept test for power tools.

(Continued)

(*Continued*)

industry, number of crews/jobsites, number of workers in crew) and on the interviewee (role in the decision-making process, position in company, number of times interviewee personally uses studs). The prototype test also included general information about the stud anchors that were used (e.g., type and quantity of the stud anchors used, purchased brands, diameters of anchors, applications for which stud anchors are used) and the stud anchors' setting process (e.g., explanations of the normal way of drilling a hole, of setting the stud anchor, tightening the nut, and checking if the stud anchor has been properly set).

The concept test showed that interviewees perceived the AQA prototype as improving safety, preventing studs from breaking loose due to the wrong torque, and improving productivity, because it reduces the time needed for setting the stud and checking the setting. The test also provided feedback on features that could be improved and on the most attractive target markets (Butzerin, 2011).

4.5 Concept tests

Concepts range from rough ideas to functional product prototypes. Concept tests occur in the early stages of the innovation process to assess the product's attractiveness from its potential users' point of view and prior to the innovation company investing substantial resources in its development. Concept tests help differentiate strong concepts from poor ones, thus helping the company choose those worth carrying forward to development and filtering out poorly received offers. In addition, concept tests provide input for segmentation strategy, as they indicate the potential future product's attractiveness for different customer segments. Further, the respondents usually explain their concept likes and dislikes during concept tests and make potential improvement suggestions. These insights help refine and improve concepts' customer value (Page and Rosenbaum, 1992; Crawford and Di Benedetto, 2008).

Irrespective of the degree of the design elaboration, respondents base their assessment of a concept on its description and presentation but have no opportunity to experience its function, because it has not yet been developed to a functional level. As soon as a functional prototype is available, product tests can take place.[2]

Customer integration in action: Security trading of concepts

Concept tests can be carried out in a studio or online. Testers increasingly try to include gamification elements to engage the respondents. The security trading of concepts is a novel, playful approach in which securities represent competing product ideas. The respondents assume the role of stock traders betting on promising ideas. They play several rounds of this security trading of concepts game and the winner receives a prize. The basic idea of this approach is that the respondents indirectly express their desires, needs, preferences, and expectations about new product ideas and future events by buying and selling securities. By observing the respondents' buying and selling transactions, the testers draw inferences about the preference for and market acceptance of the ideas. This security trading game is enjoyable and motivating for respondents, but leaves room for opportunistic behavior by the players that reveals little of their real product preferences but instead demonstrates their strategic gaming capabilities (Dahan *et al.*, 2011).

A concept test is an experiment with the concept as a stimulus. A concept describes a new product idea's characteristic elements as precisely as possible at the current ideation stage. Concept designs range from stripped verbal descriptions without visuals to virtual concepts and, finally, to nonfunctional physical prototypes. The concepts of services usually come in the form of service blueprints and describe the service process (see Section 3.6. for details). When the physical representations of a service

---

[2] In literature and practice, there are some inconsistencies in the terminology. For example, the first functional car models are called concept cars in the automobile industry. In testing terminology, these concept cars are functional prototypes that allow for product tests. The literature occasionally categorizes prototype tests as concept tests, product tests, or as a separate test category. We differentiate between concept and product in as far as concepts do not yet function, while products incorporate functions. Prototypes can be either one or the other category, depending on the availability of functional elements. The reason for this differentiation is that it allows the distinction between the two tests to be clearly identified: Concepts require respondents to imagine the possible functioning of the future product, but products allow them to test the functioning and report on the quality of the realized functioning.

are relevant, concept tests for these tangibles can be carried out in much the same way as concept testing for NPD/NSD (Tidd and Hull, 2003; Lees and Wright, 2004). For example, napcabs is a service offering cabins for sleepovers or quiet office space at airports. The concept testing in the early phases of this service development needed physical representations of the cabins to gauge customers' interest in such a service, because people associate a wide range of different things with the word "cabin." Concepts and early physical prototypes of the cabin were essential for customers to understand what they could expect from such a service (Rau *et al.*, 2016).

Stripped verbal descriptions are brief and bland descriptions of the concept. Their main benefit over more elaborated concepts is that they allow for testing a larger number of concepts quicker and cheaper. Embellished concepts incorporate emotive, persuasive formulations offered in a promotional way. Embellished concepts allow for conveying a concept's emotional elements and are appropriate where these emotional elements are a key characteristic of the concept. For example, an embellished concept for a new perfume scent may convey the perfume's communicative positioning (e.g., the way the perfume's scent shall be communicated in an advertisement) as a central denominator of its attractiveness to a certain customer segment, irrespective of the actual scent. In other cases, embellishment can bias results toward greater acceptance (Dolan, 1993; Crawford and Di Benedetto, 2008). The design of a concept, or a stimulus, is not a trivial issue, as quality and elaboration may influence purchase intention scores.

Visuals are a quick and easy way of conveying basic information about a concept, such as its size, weight, and certain aesthetic and functional aspects, which are difficult or lengthier to explain with nonvisual concept stimuli. If visuals, such as simple line drawings for familiar, low-complexity product categories like toothpaste, do not contribute shorter or additional information to the verbal stimuli (Dickinson and Wilby, 1997) or visuals only provide negligible information (e.g., color) instead of pivotal product information (e.g., taste, effect on teeth whiteness, or tooth decay), both forms of stimuli deliver similar test results. In such cases, limiting concepts to nonvisual stimuli may increase the number of concepts testable in a concept test session and decrease the cost of concept testing (Lees and Wright, 2004). Currently, however, CAD software and other IT tools allow the fast, engaging, and cost-efficient design of elaborate visuals in the form of virtual prototypes that improve decision quality, engender trust, and support respondents' comprehension and understanding of the final product, In addition, such visuals have been proven to yield realistic and roughly similar results as the tests of final, functional products (Domzal and Unger, 1985; Urban *et al.*, 1997; Dahan and Srinivasan, 2000).

It is important to note that concept test participants confronted with elaborate, embellished, and slick concept designs are inclined to accept the concept as resembling the final product and evaluate it as such. When confronted with a rudimentary design, they are inclined to understand this as an invitation to suggest changes to the concept or even to tamper with it (Thomke, 2014). This suggests that the first type of concept design is more appropriate for quantitative concept tests aimed at gathering market size and market acceptance data in later NPD/NSD phases, while the latter's value lies in its use in qualitative concept testing in the earlier NPD/NSD phases when the test participants are invited to provide improvement suggestions.

Generally speaking, concept tests are not applicable to all product categories. In concept tests, the actual features cannot yet be experienced. The function or future value can only be described. The value of such tests is limited for products whose key buying criterion is customers actual liking the way its features have been realized. Perfume is an example: A concept test describing a perfume's scent cannot test whether consumers will actually like the scent, or not, once they can really smell it. However, such a test can provide some indications of how to describe the scent in the innovating company's communication strategy (Table 18).

Table 18: Strengths and challenges of concept tests.

Strengths	Challenges
• Relatively cost- and time-effective way to obtain the first feedback on a new product idea • New opportunities for realistic customer feedback through virtual reality and other methods that increase respondents' abilities to imagine future product use	• No functional testing or real product experience possible • Data on market acceptance is based on articulated adoption intention instead of real behavior • Respondents might misinterpret, or misunderstand, the concept idea • Limited value for product categories with high experience value (e.g., perfume) • The size of the customer base in B2B markets often reduces the applicability of quantitative concept testing

4.6 Product tests

Product tests include all tests in which the potential customers evaluate a functional prototype, product, or product element. Basically, the innovating company is interested in the same set of questions as that used in concept tests. Product tests' key goals include insights to help with product optimization and to understand different customer segments' perceived benefits, their willingness to pay for market-targeting strategies, and their gauging of the market potential and the market size, as well as receiving suggestions regarding refining the target market positioning.

While concept tests are limited to explaining a product's potential function, product tests allow the participants to experience a product's functioning and the way it works, as well as its quality level, performance, and its application possibilities. The participants subsequently report on these experiences.

Product tests can be partial tests or full product tests. A full test captures perceptions of the product as a whole and its different features. Partial tests focus on a specific product feature, such as the packaging, the taste/smell, the design/form, the usability, or the brand. In such tests, the other product features are neutralized, eliminated, or kept the same in different versions to allow the participants to only focus on the selected feature (Table 19). For example, a taste test may ask the respondents to

Table 19: Strengths and challenges of product tests.

Strengths	Challenges
• Functional testing allows for detecting application insufficiencies • Respondents' feedback is based on (preliminary) experience with a new product • Qualitative product tests can be applied to any product category in the B2B and the B2C markets • Beta testing is popular and useful in B2B markets and in the software industry	• Prototypes can be expensive • Market acceptance data are based on articulated adoption intention instead of real behavior • The size of the customer base in B2B markets often reduces the applicability of quantitative product testing

Research note: The role of design elaboration and visualization techniques in concept tests for radical innovation

Mugge and Dahl (2013) carried out a series of 2 × 2 experiments in which they varied concept innovativeness in the textual description (radical, incremental) and in the *concept visualization* (*deviation/no deviation from the product category's current design state*) for various consumer product categories. Four groups of respondents randomly received one combination of textual description and visualization. Figure 39 shows washing machines as an example.

In terms of incremental concepts, visualizations that deviate from the current design state make the concepts more interesting for the respondent. The opposite holds true in terms of radical concepts: Visualizations deviating from the current design state confound respondents, thus hampering their ability to understand the concept and transfer knowledge from the existing product category. Moreover, large deviations from status quo make respondents fear the cognitive effort required to make effective

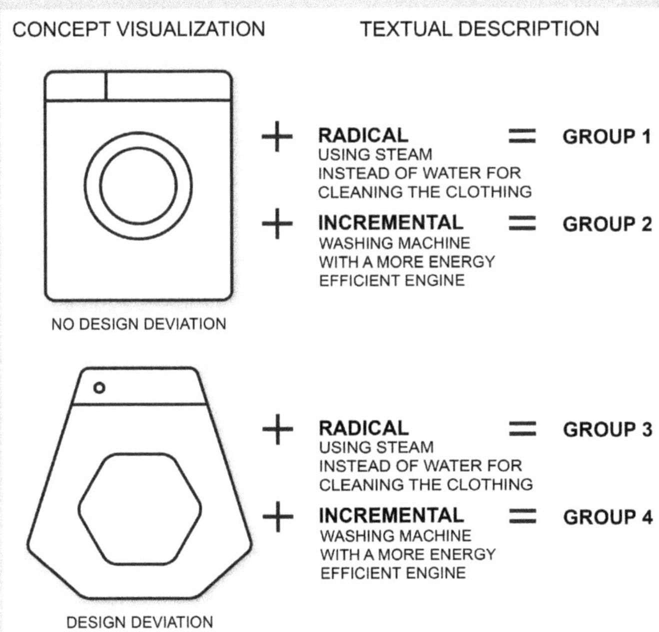

Figure 39: The role of concept visualization in radical concept tests.

(Continued)

(*Continued*)

use of a product, resulting in skepticism toward the concept. Visual stimuli used with radical product concepts that clearly allude to a specific product category help respondents categorize these concepts, resulting in more favorable concept evaluations.

However, the innovating company has to pay attention to the choice of product category to which the radical concept alludes. Categorization cues not only reduce the perceived complexity but also influence performance expectations. Moreau *et al.* (2001) found that respondents evaluated digital cameras (when the product category was still young and emerging) compared to standard cameras lower than those compared to scanners, because their color quality was then relatively better than that of scanners but worse than that of cameras.

Besides product category cues, research studies the value of multimedia presentations, mental visualization techniques, and narratives to make radical concepts more accessible for respondents participating in concept tests (Figure 40). *Information acceleration* is a technique that puts respondents into a future environment by means of multimedia simulation of the future environment, the new product, and its marketing mix. Urban and colleagues (1996, 1997) presented forecasts about radical products based on the information acceleration technique and found that the result had good internal and predictive validity. Nevertheless, the method has been applied rarely, because multimedia realization is expensive, time consuming, and customers' evaluation of a concept presented via information acceleration depends strongly on the multimedia presentation's quality and content.

Mental visualization techniques encourage respondents to visualize using the radical product and think of new activities that they can perform with the product that they have never done before. This technique unleashes respondents' imagination and makes the radical concept's value real, resulting in more favorable evaluations. A downside is that the innovating company has little influence on the direction into which the respondents' imagination moves and the new, imagined activities may not fit the radical concept (Zhao *et al.*, 2009; Zhao and Zauberman, 2011).

Concept description in a narrative format strives to combine the strengths of information acceleration and mental visualization while avoiding their weaknesses. In this approach, the respondent learns about the radical concept by reading a story of someone using the product in a future usage scenario. The innovating company enables the respondent

(*Continued*)

(*Continued*)

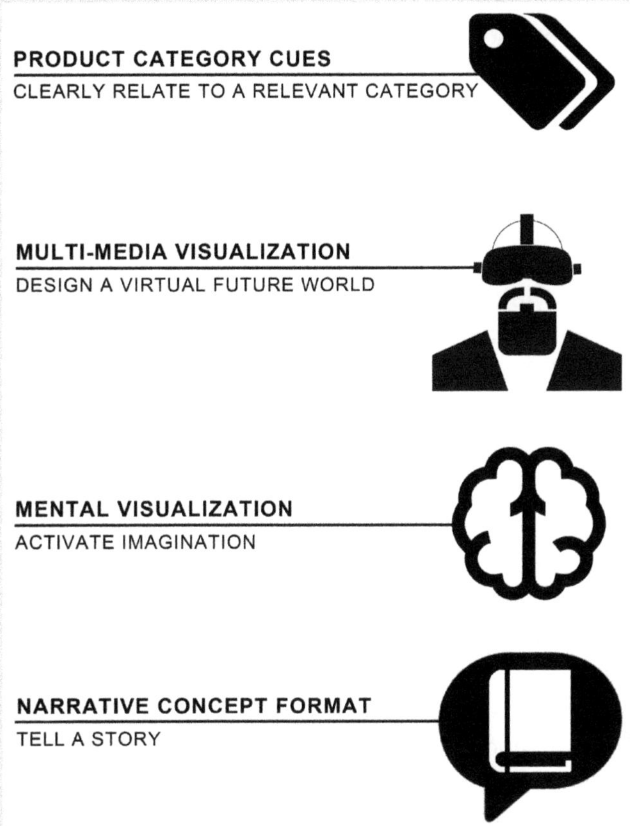

Figure 40: Methods to improve respondents' ability to imagine future radical product usage in concept test settings.

to construct vivid visual images of specific activities that a customer can actually perform with the new product. Such concepts can be realized relatively cheaply and quickly and increase respondents' imagining of the benefits, their concept evaluations, and their ability to provide meaningful improvement suggestions (van den Hende *et al.*, 2012; Schweitzer and van den Hende, 2017)

See the following video for more on the narrative concept format: https://www.youtube.com/watch?v=ecyDzKVo5z0 (accessed 30 August 2017).

drink four different orange juices. The glass in which the juice is presented is unbranded, and the same one is used for each juice. After tasting, the respondents provide feedback on their overall preference, the perceived sweetness, smoothness, and other sensory perceptions.

Product tests can take the form of short-term studio tests and long-term home/office/factory tests. In the short-term version, participants visit the site of the innovating company, or an involved market research agency, to participate in a product test that may take half an hour to half a day. Studio tests provide insights into spontaneous reactions to and associations with an offering, while home tests provide better insights into customers' usage patterns, ways of applying a new offering, and the offerings benefits for them and the difficulties they may have with it. The cost per respondent of home tests of a new toothpaste, beauty cream, or detergent is quite small; respondents are easy to find; and they can easily take the product home to try it over several weeks. In B2B markets for complex and costly equipment, an innovating company may find identifying a feasible sample size for quantitative testing challenging. It might be difficult to justify shipping a new €280,000 production machine to 400 organizations for in-house tests. Consequently, such tests generally are done in-house, or at trade fairs. Qualitative home tests are in such cases only undertaken at a very late stage in the form of beta testing (Berekoven *et al.*, 2004).

Beta testing comprises a late-stage product test. The term beta testing originated in the computer industry and refers to a selection of potential adopters testing the functions of new software in their real-world environments. Conversely, alpha tests describe the functional tests that an innovating company's R&D undertakes to simulate customer use. Beta testing starts once the alpha testing has been completed and the known bugs have been solved. The beta test strives to find bugs that cannot be anticipated, or realistically simulated, in alpha testing. Bata testers report problems they identify to the R&D team as soon as they detect them, as well as providing the team with weekly feedback via interim status reports and other formal and informal ways. Besides checking the functions of a new product, beta testing also has promotional and forecasting goals. For example, prospective customers might be allowed to visit the beta-test site to experience the product in action; prestigious beta testers may serve as references to legitimize the offering; the prospective customers' uncertainty about the offering could decrease if publicity about successful tests spreads; contact with key customers during beta tests could improve and foster customer relationships; after the beta-testing period, the beta tester could be offered the test product at a reduced price; and a comparison of the beta testers who actually buy the beta product with those returning it could serve as a good indicator of the market acceptance. Nevertheless, beta testing can also backfire if an offering does not live up to customers' expectations and engenders negative word of mouth and publicity (Dolan and Matthews, 1993; Jiang *et al.*, 2017).

Customer integration in action: The value of car clinics as industry-specific product tests

A car clinic is a product test in the automobile industry. Respondents try out one or more cars, report their perceptions, and evaluate the tested cars. Car clinics provide insights into the tested cars' strengths and weaknesses, as well as being a helpful tool for assessing cars' market acceptance and for making sale forecasts. Static car clinics are tests done in an indoor space. These tests often focus on the exterior and interior design aspects and include sophisticated multisensory methods to measure the emotional impressions that certain features leave on respondents (e.g., emotions aroused by haptic experiences such as touching the seat material). Useful tools include eye-spying webcams combined with intelligent image-processing algorithms to analyze respondents' facial expressions for positive and negative feelings. Dynamic car clinics include road trials and provide either overall feedback on the car, or feedback on specific features, such as a car's automatic parking assistants. In these trials, the respondents perform various tasks, for instance, parking the car in easier and more difficult parking spots. In this case, the car manufacturers are interested in feedback on, for example, the understandability of the system, the human–machine interface, a preference for either visual or acoustic warnings, ease of use, willingness to pay, and adoption intention.

Car clinics include a mix of qualitative and quantitative survey elements before, during, and after the actual testing of the car. The respondents are usually recruited from the current customers in the audience at whom the innovation is targeted (e.g., the current drivers of upper-class cars).

(Continued)

(*Continued*)

Sometimes car clinics are undertaken without any brand identification. This is especially beneficial if these clinics focus on the functionality and the perceived relevance of certain features. If an innovating company wants to study the effect of certain design aspects, branded clinics are more adequate, because a design's fit with a specific brand image is highly relevant in the car industry.

Traditionally, car manufacturers used car clinics rather later in the NPD process and mainly to gain insights into details regarding their positioning strategy and to identify the handling problems users might experience with a car. Currently, manufacturers and suppliers increasingly strive to adapt car clinics to allow earlier customer feedback. This endeavor focuses on concept testing with virtual reality representations to demonstrate what certain functions will later look like and how they will work. For example, the US car manufacturer Ford uses Wii (a game console) remote controls to show respondents a car concept; the respondents can then examine this in such detail that they can even open the virtual glove compartment or move the virtual steering wheel.

Generally speaking, product tests provide good feedback on a product's functioning, its perceived benefits, and its value for different potential target groups. The product tests also allow the manufacturer to gauge the product's market acceptance and to forecast both its market potential and its sales potential. Since considerable resources have already been invested in a product when it is ready for product testing, the sunk costs are high if the market response is negative and profound changes to the product features delay the time to market. Consequently, later stage tests usually only focus on avoiding teething problems and glaring errors. The innovating company thus only uses radical improvement suggestions as input for the next product generation, instead of incorporating them into the current development project. Conversely, early stage prototype tests are used to challenge the solution, to select the most promising features, and to improve the prototype's overall quality.

Product tests can be monadic or nonmonadic. Monadic tests only test a single product. Sequential or parallel tests are nonmonadic tests in which respondents evaluate several products during a session either consecutively or simultaneously. The products compared in nonmonadic tests can be either different prototypes that the company plans to launch or current competitive offerings. Order effects are a problem in nonmonadic testing, which means items presented first tend to be rated higher than the rest and also serve as reference points, with subsequent offerings being evaluated overly negative if they follow good ones and overly positive if they follow bad ones (Friedman and Schillewaert, 2012). Furthermore, respondents who have to evaluate a large amount of products become tired and bored, thus evaluating later offers less critically and increasingly accepting the presented offers. Although these order effects seem to dispute nonmonadic testing, such tests have clear cost and time advantages over monadic ones. Randomly varying the position in which the different products are presented is a way of reducing the methodological biases of order effects.

4.7 Conjoint tests

The conjoint test is a quantitative testing method that can use either concepts or products as stimuli. Differing from the traditional concept and product tests presented in the previous sections, the conjoint measurement technique is a decompositional technique. Decompositional refers to the special way in which the concept or product stimuli are presented. In traditional techniques, respondents evaluate the attractiveness of certain product features separately. For example, customers evaluate a list of attributes, such as the relevance of the brand, processor capacity, RAM size, panel display size, and price on a scale ranging from very important to not important at all (see Figure 41). These questions are supplemented by a number of ancillary questions on the preferred brand name, the ideal MB for the processor, and so on. The problem with this form of questioning is that the respondents might rate all these attributes as highly important. The ancillary questions might show that the respondents, in a range of options, consider the laptops with the highest MB, the highest GB, the biggest size, and the lowest price as the preferred ones, leaving the innovating company with little relevant information with which to develop a new product further.

The idea behind the decompositional technique is that customers usually do not need to decide between individual attributes, which can be varied at will, but that they buy product bundles, demonstrating a

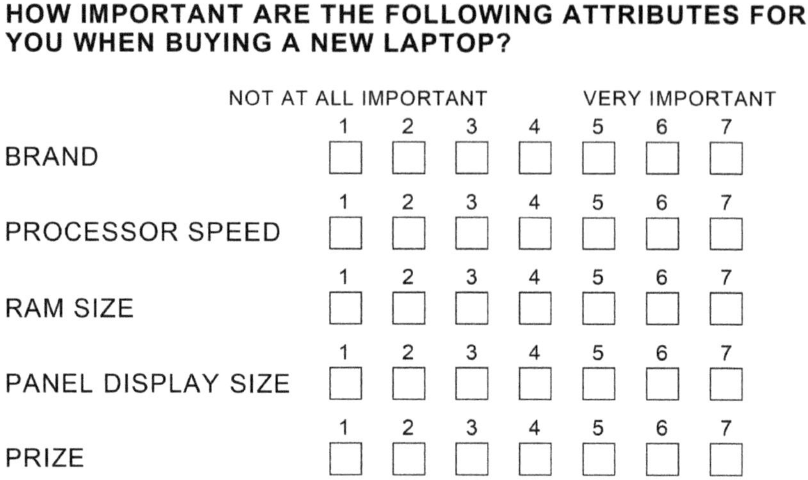

Figure 41: Traditional question format in quantitative concept/product tests.

preference for a product bundle representing the added preferences in respect of each attribute. When buying products, customers have to make trade-off decisions between certain quality levels and the price levels. Similar to such real-life situations, the decompositional technique confronts the respondents with product bundles comprising sets of total profile descriptions consciously created from different attribute options (e.g., processor speed, RAM size, battery, price) and attribute levels (e.g., 3.1 GHz, 3.4 GHz, 3.8 GHz cache for processor speed; 16 GB, 18 GB, 20 GB for RAM; 10 h, 12 h, 14 h for battery life; €899, €960, €1,150 for price). Figure 42 presents an excerpt of the total profile descriptions in a conjoint survey.

Based on the evaluation of the different product bundles, or in conjoint terms, product profiles, the innovating company can calculate the attributes' utility functions by decomposing the full bundles into each attribute's contribution (Green and Srinivasan, 1990). A conjoint test of laptops with two attributes and two and three attribute levels, say a 3.4 GHz/3.8 GHz processor speed and a 10 h/12 h/14 h battery life, would have to evaluate six bundles in a full factorial design (3.4 GHz/10 h, 3.4 GHz/12 h, 3.4 GHz/14 h, 3.8 GHz/10 h, 3.8 GHz/12 h, 3.8 GHz/14 h)

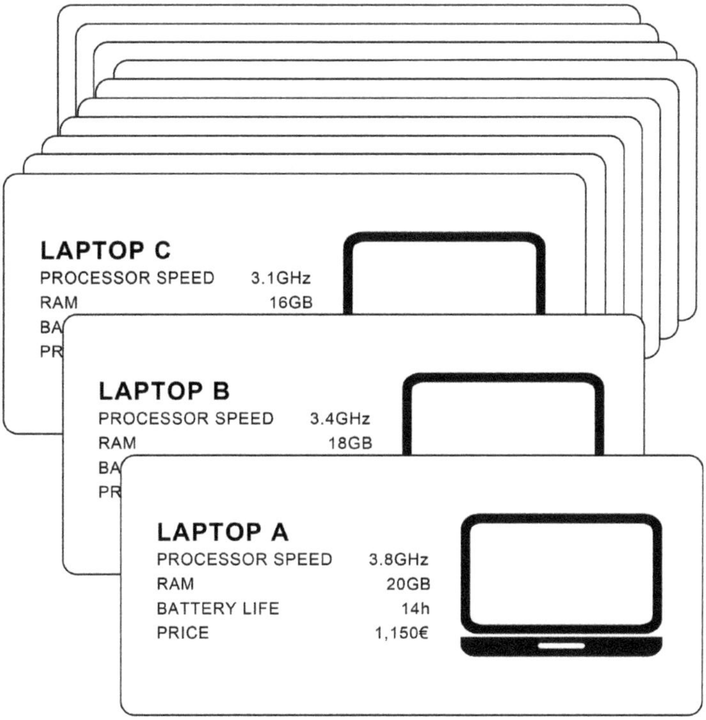

In the following you find 12 cards with descriptions of new laptops (Model A to L) that vary in terms of processor speed, RAM size, battery life and price. Please study them thoroughly and rank-order all of them by assigning twelve points to what you consider to be the best laptop, eleven points to the 2nd best laptop, ten points to the 3rd best laptop and so on ending with one point to the least preferred laptop.
Enter your ranking in the box below:

Nr.1 (=BEST LAPTOP 12 POINTS)	Nr.2 (11 pts)	Nr.3 (10 pts)	Nr.4 (9 pts)	Nr.5 (8 pts)	Nr.6 (7 pts)	Nr.7 (6 pts)	Nr.8 (5 pts)	Nr.9 (4 pts)	Nr.10 (3 pts)	Nr.11 (2 pts)	Nr.12 (=LEAST PREFERRED. 1 POINT)

Figure 42: Conjoint question format.

by ranking the bundles, starting with 12 as the most preferred bundle and ending with 1 as the least preferred bundle. Table 20 shows the preference ratings of an exemplary respondent. While it is not at all surprising that a respondent would prefer the 3.8 GHz/14 h version most and the

Table 20: Individual answer pattern in a conjoint study.

Attributes	Battery life			
	Attribute levels	10 h	12 h	14 h
Processor speed	3.4 GHz	1	3	5
	3.8 GHz	2	4	6

3.4 GHz/10 h version least, the ranking of the remaining product bundles provides insights into the attributes' perceived relative importance for a respondent. In Table 20, the respondent assigns more importance to battery life than to processor speed. Based on the respondent's preferences, the innovating company then calculates the part-worth utilities of all the attribute levels.

In other words, the decompositional technique assumes that consumers evaluate product bundles by evaluating the separate, yet conjoined, parts and adding these partial evaluations to build their preferences for the full product bundle. Since the estimated overall utility of, or preference for, a product bundle is equal to the sum of its parts, the part worth of particular attributes and attribute levels can be calculated on the basis of a set of full product bundle evaluations.

This way of gathering data allows the innovating company to gather insights into the relative importance of attributes; for example, customers are willing to trade off a shorter battery life for a higher processor cache. Conjoint settings also allow for scrutinizing the relevance of price and customers' price sensitivity. This information helps the focal innovating company understand on which attributes and attribute levels it should focus in order to build a better laptop.

Conjoint analyses follow a strict sequence of steps, which Figure 43 depicts. First, the attributes and the attribute levels require specification. Since only a limited number can be included, the choice of those that could potentially have a strong influence on buying behavior, and which the innovating company can influence and realize, is pivotal for a meaningful output. A prestudy is advisable to make the right choice. To illustrate the process in detail, consider a conjoint setting for a new laptop with three attributes and three attribute levels each (processor

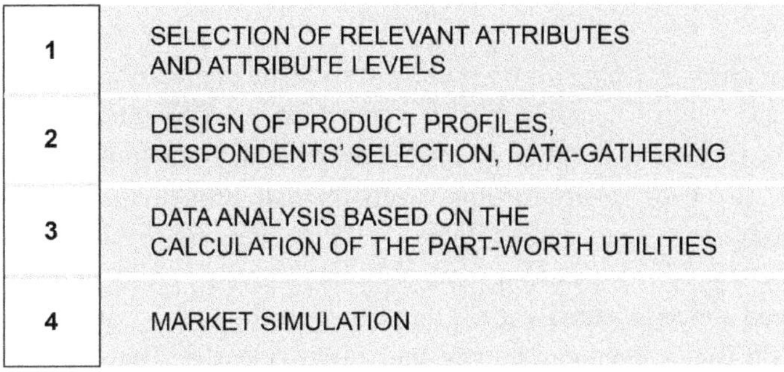

Figure 43: Steps in carrying out a conjoint analysis.

speed: 3.1 GHz, 3.4 GHz, 3.8 GHz; battery life: 10 h, 12 h, 14 h; price: €899, €960, €1,150).

Next, the innovating company specifies a conjoint measurement technique (see the box *Customer integration in action*: *Choosing the correct conjoint technique* in Section 4.7 for the different options), sets up the study, specifies the target group and the sampling technique, and collects the data.

Subsequently, individual part-worth utilities are calculated and aggregated for each respondent. The key principles for these steps are shown in Figure 45: Based on the respondent' preference rankings, the means of each attribute level and the average value of these means are

Customer integration in action: Choosing the correct conjoint technique

Consider a company interested not only in two attribute levels for the processor capacity and three for the battery life but also in four for the brand, two for the graphic board, two for the Webcam quality, three for the weight, four for the RAM, and three for the price. It is easy to see that possible combinations explode with every inclusion of another attribute ($2 \times 3 \times 4 \times 2 \times 2 \times 3 \times 4 \times 3 = 3{,}456$ possible profiles in this case). By systematically varying the attributes and attribute levels of the profile design by means of mathematical algorithms, this reduces the number

(Continued)

(*Continued*)

of alternatives that have to be presented to respondents to a fraction. The part-worth scores calculated from the presented set of product profiles then allow for predicting the respondents' reaction to all 3,456 possible product profiles. However, this reduction affects the data reliability and limits the manageable set of attributes (Huber, 2005; Netzer and Srinivasan, 2011). Over the last few decades, several conjoint analysis variations have been developed aimed at increasing the number of attributes that can be included, thus reducing respondent fatigue and improving the data accuracy and consistency. Many of these conjoint analysis variations are software based. Sawtooth is one of the leading providers of relevant software:

Traditional and adaptive conjoint analysis (*ACA*). In the traditional conjoint test, respondents sort the presented product profiles from the most to the least preferred version. This method is the most cognitively strenuous approach for the respondents, as they need to rank a considerable number of profiles. This is the only not mandatorily IT-based conjoint version and thus the only one with a paper-and-pencil style. The adaptive conjoint analysis is a software-based version of the traditional approach but dynamically adapted to a respondent's previous answers, thus allowing the adaptation of the conjoint design to the individual preferences of the respondent in turn resulting in the possibility to include up to 30 attributes. These conjoint versions are mainly used for feature choice decisions in NPD and can only be used with small sample sizes (smaller than 100).

Choice-based conjoint (*CBC*) *analysis*. In choice-based settings, respondents are repeatedly asked to choose between different sets of product profile options. These options can include the option not to choose any of the proposed profiles (see Figure 44). Compared to traditional conjoint settings, choice-based conjoint analysis reduces respondents' cognitive evaluation effort. Choice-based designs simultaneously increase the realism of the conjoint setting, as consumers usually compare a limited set of alternatives and may as well not choose any option at all. The traditional conjoint setting does not take the latter into account. A disadvantage of this conjoint version is that respondents repeatedly only choose the most preferred version of a set of options. In other words, the extent to which the respondents prefer this option to other options is not known. Another downside is that respondents

(*Continued*)

(Continued)

IF YOU WERE LOOKING FOR A NEW LAPTOP, AND THESE WERE YOUR OPTIONS, WHICH WOULD YOU CHOOSE?

LAPTOP HP

PROCESSOR SPEED	3.8GHz
RAM	20GB
BATTERY	14h
PRICE	1,150€

LAPTOP LENOVO

PROCESSOR SPEED	3.4GHz
RAM	18GB
BATTERY LIFE	12h
PRICE	960€

LAPTOP DELL

PROCESSOR SPEED	3.1GHz
RAM	16GB
BATTERY LIFE	10H
PRICE	899€

NONE: I WOULDN'T CHOOSE ANY OF THESE.

Figure 44: CBC question format.

find the repetitive choosing of their preferred options from rather similar-looking sets of options a tedious task. Choice-based conjoint analysis is currently the most widely used conjoint version, particularly for pricing decisions.

Partial-profile CBC analysis. Similar to CBC, respondents choose a profile from a set of profiles, but these profiles are subsets of a limited number of attributes instead of full profiles with all the attributes. The individual-level stability is lower than that in other techniques, because each task has

(Continued)

(Continued)

substantial attribute omissions. However, this technique can deliver good market simulation results if the sample size is large enough.

Adaptive CBC (ACBC) analysis. This conjoint technique mixes ACA and CBC. The respondents first configure a product offering that represents their ideal product version. Thereafter, the software builds on this information, constructing 10 full profiles that the respondents have to evaluate in a CBC setting. ACBC tests usually take longer than CBC tests, although respondents perceive ACBCs as more engaging, because the concepts are closer to their preferences. While CBCs work better with four or less attributes, ACBCs are the right options for five and more attributes. Similar to CBC, ACBC is better than other techniques for pricing studies.

Menu-based choice (MBC) analysis. This approach is suitable for products that customers usually buy in menu format (e.g., computers, cars). Menu formats enable respondents to assemble the final product themselves by choosing from different options. The MBC analysis mimics such configuration options (Orme, 2013).

calculated (step 1). The part-worth value of each attribute level is the difference between the average value and the mean of the attribute level (step 2). Next, the rank values are derived, which represent the difference between the highest and the lowest part-worth values; the highest part-worth value of the attribute levels within an attribute is also calculated and the range values of all the attributes are summarized (step 3). By dividing the range of the attributes by the summarized range of all the attributes, the innovating company can determine an attribute's relative importance (step 4).

The respondent whose data are presented in Figure 45 attributes the highest relative importance to battery life, while price is the least important for him. Based on this information, the product bundle with the highest utility for this respondent can be easily identified (14 h battery life, 1.6 kg weight, 3.8 GHz processor speed, €899 price). The total utility that the respondent derives from this product bundle is 30.52. Similarly, the utilities of any attribute level combination can be calculated on the basis of the respective part-worth utilities. Table 21 depicts possible utilities

		STEP 1		STEP 2	STEP 3	STEP 4
Attributes	Attribute Levels	Mean of Attribute Levels	Mean of Attribute	Utility Scores PWU	Range	Relative Importance RI
Battery Life	14h	9.50		*3.00*		
	12h	7.25		*0.75*		
	10h	2.75	*6.50*	*-3.75*	*6.75*	*0.46*
Weight	1.6 kg	7.88		*2.00*		
	2.4 kg	6.27		*0.39*		
	3.2 kg	3.49	*5.88*	*-2.39*	*4.39*	*0.30*
Processor Speed	3.8 GHz	7.01		*0.68*		
	3.4 GHz	6.85		*0.52*		
	3.1 GHz	5.13	*6.33*	*-1.20*	*1.88*	*0.13*
Price	€ 899	7.85		*0.86*		
	€1,150	6.13	*6.99*	*-0.86*	*1.72*	*0.12*
TOTAL		*(mean)*	*6.43*	*(sum)*	*14.74*	

Figure 45: Individual part-worth calculation of respondent 1.

Table 21: Individual answer pattern in a conjoint study.

	Battery life	Weight	Processor speed	Price	Utility
New product option 1	14 h	1.6 kg	3.4 GHz	1.150 €	—
	3.00	2.00	0.52	0	5.52
Competitor A	12 h	2.4 kg	3.8 GHz	1.099 €	—
	0.75	0.39	0.68	0.35	2.17
Competitor B	12 h	3.2 kg	3.1 GHz	970 €	—
	0.75	−2.39	−1.20	1.23	−1.61

for a new product and competitive products, showing that respondent 1 attributes a higher utility to new product option 1 than to the other two options. It also shows that respondent 1 prefers the option of competitor A to that of competitor B. Note that the different price levels between the minimum and the maximum price point are calculated with the following formula to determine the utility values: € P_i = (MaxPrice – P_i) × (Range)/(MaxPrice – MinPrice).

In the next step, the innovating company may undertake what-if market simulations. Through these simulations, the company can determine the percentage of customers preferring a certain new product compared to competitive products. Table 21 provides the utilities that different respondents derive from three product options: an intended new product launch and two competitive offers. Assuming a customer is most likely to choose the product bundle providing the highest utility, a logit model calculates the market shares of different product bundles with

$$\text{Share}_i = \exp(\text{c} \times U_i) \Big/ \sum_{j=1}^{n} \exp(\text{c} \times U_j),$$

where

U_i = utility of product bundle i
U_j = utility of product bundle j
c = confidence parameter
n = total number of product bundles in the competitive set, including product bundle i.

Using this formula to calculate the market shares based on the new product offering's attractiveness for respondent 1, with the logit as a just ratio of attractiveness to the total attractiveness of the market offerings, leads to a forecasted share of 41.7%.

Running several simulations with different features for the new product profile, while holding the competitive offerings constant, provides insights into the market preferences under each condition. This also identifies the competitors from which a new product will take the most share if a certain attribute level is increased or decreased. Such simulations also provide insights into new products' possible cannibalization effects, for example, due to the launch of a new product with lower quality attribute levels. In this case, simulations provide information on the likelihood of current customers of a higher-end product version switching to the new product instead. The simulations also facilitate the calculation of different product adoption likelihood for different segments (i.e., group of people with homogenous attribute preferences that differ from other group's

Table 22: Market simulations with individual persons.

	New product option 1	Competitor A	Competitor B
Person 1	5.520	2.169	0.113
Person 2	4.400	5.310	1.280
Person 3	3.790	6.820	2.440
Person 4	−1.250	−1.080	4.730
Person 5	7.500	4.900	2.220
	$c \times U$		
Person 1	0.458	0.180	0.009
Person 2	0.365	0.441	0.106
Person 3	0.315	0.566	0.203
Person 4	−0.104	−0.090	0.393
Person 5	0.623	0.407	0.184
	$\exp(c \times U)$		
Person 1	1.581	1.197	1.009
Person 2	1.441	1.554	1.112
Person 3	1.370	1.761	1.224
Person 4	0.901	0.914	1.481
Person 5	1.864	1.502	1.202
	Choice probability		
Person 1	0.417	0.316	0.266
Person 2	0.351	0.378	0.271
Person 3	0.314	0.404	0.281
Person 4	0.273	0.277	0.449
Person 5	0.408	0.329	0.263

attribute preferences). The innovating company can also run models that include its competitors' reactions (e.g., competitor A decreases the price of its offer by 10%). These simulations provide information that help companies find the optimal product configuration for a new product while taking the company's and its competitors' current market offerings into account (Gustafsson *et al.*, 2007) (Table 22).

The decompositional technique is closer to real choice situations than other techniques and forces respondents to take trade-offs into consideration. Consequently, a conjoint analysis is likely to provide more realistic importance weights for different attributes and attribute levels than traditional concept/product tests.

We have mainly focused on describing the basic functioning and the key strengths of conjoint tests, but they also have their downsides. Similar to other quantitative tests, they provide no opportunity to identify new, potentially important, attributes that the innovating company has not yet considered. Put differently, conjoint tests build on predetermined features that prior research identified as important for similar product categories; hence, it is difficult to capture the value of innovative features that customers do not yet consider in their product evaluations. Moreover, the high task complexity of such evaluations exhausts respondents, causing respondent fatigue. This fatigue could potentially result in respondents dropping out of a test, or in being biased, because they only focus on one or two attributes of the full profiles to complete the task as quickly as possible. Last but not least, the detailed way of comparing feature is not quite adequate for certain product categories, such as low-involvement products, products whose consumption decisions consumers make spontaneously, impulsively, and emotionally, as well as products merely bought for prestigious reasons. When wishing to purchase such product categories, customers do not make intensive attribute-based comparisons. In other words, the sophisticated method of collecting respondents' perceived quality expectations by means of conjoint measurement techniques is not very widely used. Product categories that involve rational, complex decision processes, such as high-involvement products, which include many B2B products, are better candidates for conjoint tests (Davis, 1993; Dahan and Srinivasan, 2000; Orme, 2013) (Table 23).

4.8 Store and market tests

A store test is one that takes place in selected stores that are appropriate distribution channels for a product. This implies that store tests can only be done for products that are actually sold via stores, which means the

Table 23: Strengths and challenges of conjoint tests.

Strengths	Challenges
• Allows for measuring the importance of each attribute and the degree of price sensitivity • Sophisticated calculation of the different attribute levels' utility and of the willingness to pay • The trade-off situations reflect real choice situations • Highly applicable to products involving extensive purchase decision processes	• Tiring questioning technique for respondent • Not applicable to all product categories; the respondent has to be fairly involved in the category to make meaningful choices • The application is limited to a relatively small number of predetermined attributes and attribute levels • Requires rather highly involved customers

method does not apply to many B2B solutions. The test takes the product and the usual assortment into consideration. Store tests investigate the effect of a product's whole marketing mix shortly before it is launched. At his late NPD/NSD stage, tests' main goal is to forecast the sales, the probability of a first and repeated purchases of a product, and its adoption rate as accurately as possible to ensure the product is distributed in adequate quantities, which will prevent a stock surplus and empty shelves. Besides determining a product's market acceptance, store tests can also be used to test the effect of alternative options of the marketing-mix elements, which will determine the best marketing-mix combination for the product launch. A market research agency first encourages 14–45 shops to participate in a trial sale, carries out some premeasurements, then runs the store test (with different product placement, promotional, and price options) for a period of 3–4 months before interviewing consumers and trade agents.

Market tests have the same goals as store tests and can be understood as a last rehearsal of the whole marketing mix before the actual market launch. While store tests are limited to a couple of stores, market tests are undertaken in a larger area — a whole city, region, or country. A key challenge of these tests lies in finding a test market that is representative of a large segment of the target market. In addition, if mass market communication is planned for market tests, the area also needs to be self-contained regarding media coverage. Only a couple of areas meet these representativeness and self-containing media communication criteria. The USA cities Charlotte, Indianopolis, Kansas City, Nashville, and Stockton are regularly used to test fast-moving consumer goods. In Europe, the cities Hassloch, Bad Kreuznach, Reutlingen, and Buxtehude in Germany, as well as Le Mans and Angers in France serve as test markets. Sometimes, a whole country, for example, Switzerland, is used as a test market for Europe. All these options are certainly not flawless regarding representativeness. Further, competitors know the potential test market areas well and their sales force is usually well trained in spotting new products. The competition can thus easily follow test markets' activities, gather insights into the innovating company's product strategies, hinder the tests and bias the results by running special POS activities while the innovating company carries out its test market phase, or learn from the innovating company's activities and pre-empt them with a fast copycat product that immediately addresses the total market. In addition, test markets are rather time consuming and expensive (Table 24).

Nonetheless, with regard to forecasting market acceptance, the key advantage of store and market tests over concept and product tests is that the former gather data on consumers' actual adoption behavior and the latter data on consumers' articulated adoption intention (Schweitzer, 2014).

Table 24: Strengths and challenges of store and market tests.

Strengths	Challenges
• Test the whole marketing mix • Provide good insights into the ultimate market acceptance • Provide data on customers' actual new product adoption behavior	• Expensive and time-consuming • The competition can observe the tests and interfere • The area's representativeness regarding the target population may be problematic • Store tests are limited to products actually sold in shops • Limited relevance for B2B markets

4.9 Go/no go decisions and market demand forecasts based on test results

Facilitating the go/no go decisions in respect of product concepts is one of the central reasons for carrying out qualitative and quantitative concept tests in industrial practice. While qualitative concept tests can only provide soft indications of a concept's possible popularity and attractiveness for different market segments, quantitative ones allow more accurate probability calculations.

Let's use an example to examine the two concept test types' forecasting value. Imagine a focus group with 10 potential customers of whom seven show interest in a new concept "A" and a quantitative concept test of 680 customers of whom 410 show interest in concept "B." At a first glance, you might assume that concept "A" is more attractive, because 70% of the target potential might buy the concept once it is on the market, while the percentage for concept "B" is only 60%. Nevertheless, percentage-based decisions regarding small samples are highly misleading as they do not provide accurate data on the true values in the target population. Using the statistical margin of error calculation for the proportion at a confidence level of 95% would provide a value of ±28.4% for concept "A" and of ±3.68% for concept "B":

$$e = \pm z \times \sqrt{\frac{p(1-p)}{n}} = \pm 1.96 \times \sqrt{\frac{60 \times 40}{410}} = \pm 3.68\%,$$

where p is the sample proportion, n is the sample size, and z is the z value of the level of confidence.

This signifies that the true value in the target population fluctuates between 41.6% and 98.4%. Predictions with such broad deviation margins are of little value for decision-making. Frankly, qualitative concept tests, or beta tests, should not be used as a basis for go/no go decision-making but should rather be used as a way to understand what customers appreciate, what problems they have with an offering, and how the offering, or its next generation versions, can be improved. By catering to a limited number of big customers and, as a consequence, unable to run quantitative testing, B2B companies could, as alternative measures, develop the concept further by means of preorders based on concept presentations, or readiness to invest in joint R&D. These measures might provide more accurate sales potential estimations than nonbinding statements of adoption intention in qualitative settings.

In addition to the overall adoption intention values, concept and product testing give the innovating company insights into a concept or product's relevance for different target groups. Imagine that product "B" is a radio control system for machine-to-human interaction that industries as diverse as transportation, building, mechanical engineering, and paper manufacturing might find attractive. Testing the concept in each potential target market and relating the results to the size of each target market will help the innovating company select and address the most attractive one. In the exemplary data provided in Table 25, the innovating company calculates the demand in the different target markets based on the formula in Figure 46.

The size of the relevant market is estimated by multiplying the number of companies in the industry by the average annual demand for units. While data on the number of companies need to be gathered through secondary research (e.g., from data provided by the Statistical Office of the European Union, or EUROSTAT), data on the average annual demand are best gathered by an appropriate question in the concept test. The concept test results represent the number of respondents from the specific target market who indicate a 5 or 4 on a scale ranging from 5 = "will definitely buy" to 1 = "will definitely not buy." In the mechanical engineering industry, 43% indicated a 5 and 22% indicated a 4, leading to a concept test result of 65% or 0.65. Not all respondents who indicate their buying intention actually become aware of the product and try it once it has been launched, which means that concept test results need to be adjusted. Companies that regularly carry out concept tests in a specific target market are well advised to relate past concept test results to past adoption data to construct an adjustment factor. Multiplying the target market size, the concept test results, and the adjustment factors of each industry produces the market demand data. In respect of the radio control system, the market demand data indicate that the company should focus its marketing activities on the transportation industry (Mahajan *et al.*, 2000; Kahn, 2006).

Table 25: Market demand calculation based on concept test results.

Industry	Number of companies in industry	Average annual demand for radio control concept	Size of relevant target market	Concept test results	Adjustment factor	Market demand
Transportation	7,400	78	577,200	0.73	0.52	219,105
Building	34,227	21	718,767	0.52	0.36	134,553
Mechanical engineering	1,349	35	47,215	0.65	0.43	13,197
Paper	202	42	8,484	0.43	0.64	2,335

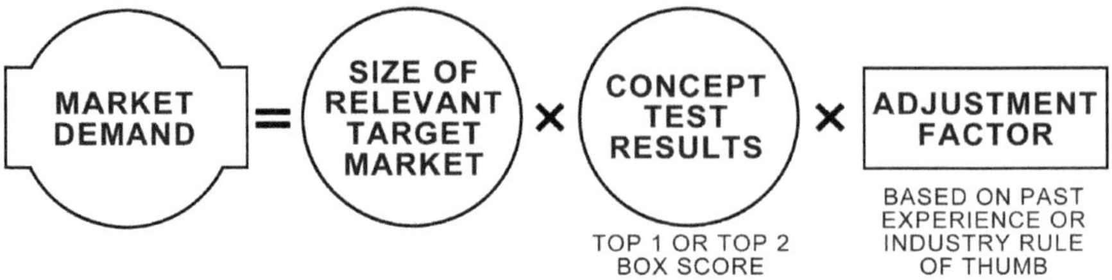

Figure 46: Estimation of new product demand based on concept test results.

The innovating company can also calculate the market penetration data based on the concept test results. The concept test results and the adjustment factor of the mechanical engineering industry indicate that 27.95% of the companies in this industry will sooner or later try the product. Based on its past experiences and its planned marketing activities, the innovating company could assume that 25% of mechanical engineering companies that have not yet tried the product will do so in any given time period t. The diffusion rate of the first three time rates can then be calculated with the Fourt and Woodlock (1960) formula: $q_t = r \times q \times (1 - r)^{t-1}$, where q_t is the number of companies likely to try the new product in period t, r is the diffusion rate, and q is the total number of companies that will try the product some time. Deploying the numbers in the formula leads to a q_1 of 0.070 equaling 922 radio control systems in the first year, a q_2 of 0.050 (or a total demand for 1,586 systems) in the second year, and a q_3 of 0.036 (2,065 systems) in the third year.

Combining quantitative concept test data and market size data help identify the most promising target markets for an innovation and assess its likely success in the market. However, other factors also need to be taken into consideration in go/no go decision-making. Put differently, when an innovating company carries out several concept tests for innovations that compete for R&D investment, it is well advised not to base its go/no go decision only on respondents' adoption intention by, for example, providing funds for the concept with the highest relative adoption intention in a concept test. Whatever the case, a certain minimum adoption intention level should be reached, as it is unlikely that a concept interesting only for, say, 20% of the potential customers in a concept test will be successful. This threshold level may vary from industry to industry and company to company. Comparing current with past concept test results can be an indication of the likelihood that the concept will succeed. If past products with an adoption intention of 60% were a success, while those with an adoption intention of 50% failed, the innovating company should determine the level of the adoption intention in a concept test at, at least, 60% in order for a product to receive a go decision at the next NPD/NSD gate. If the adoption intention surpasses the relevant threshold level, the following factors also need to be considered in go/no go decisions:

- the innovation's strategic relevance for the company
- the product innovativeness and technical feasibility
- the targeted market or market segment's specifics.

In term of a product's strategic relevance, one company might prefer a concept that provides access to a specific new market, while another prefers a concept that helps defend its position in an existing market. The market size, expected market growth, average product life cycles, competitors' actions, profit expectations, and the innovating company's current market share of specific markets also play a role when choosing the right concepts for the NPD/NSD process.

In terms of product innovativeness and a product's technical feasibility, a more risk-averse company may decide to select concepts that represent incremental product line innovations rather than more radical ones, because the latter's technical feasibility and potential market success are more uncertain. Other innovating companies may look for a good balance between concepts that are either more radical or more incremental, while still others may take chances with risky, but potentially high, returns and prefer radical concepts.

In terms of the targeted market or market segment's specifics, the above market demand calculation example shows that the relation between those customers that indicate they are interest in a product in a concept test and those who actually buy such a product might vary between markets. A possible reason for such differences is an innovating company's different access to these markets, such as a higher versus a lower distribution rate or large versus little personal knowledge of potential customers.

Chapter 5

Integrated approaches to open and close the solution space: Multiple customer interaction

5.1 Design thinking

Design thinking is neither a specific methodology nor a tool but a broader approach to product and service development. Unlike more traditional methods, which are based on data and measurement, design thinking places more emphasis on trying to understand users' actual behavior and emotional needs, with a strong focus on empathy and cocreation.

Major proponents of design thinking include the Hasso-Plattner Institute of Design at Stanford University and the international design consultancy IDEO. There are many different variations of the design thinking process model, which range from three to five stages. However, the different approaches' essence usually includes a core set of three tasks and supporting tools (Table 26).

Some design methods are targeted at reframing a design challenge, such as creative problem solving, while others are more about trying to understand users and bringing their perspective into the design articulation with, for example, techniques such as ethnography, empathic design, and construct elicitation. A more tangible element of design thinking involves creating boundary objects where user perspectives interact with

Table 26: Tools to support design thinking. (Adapted with permission from Bessant and Maher (2009). Copyright © 2009 World Scientific Publishing Company Co. Pte. Ltd.)

Design task	Tools
Reframing	Creative problem solving
	TRIZ
	Lateral thinking
	Scenarios and futures
Engaging users	Ethnographic approaches
	Empathic design
	Construct-theory-based methods
	Extreme user methods
Experimenting	Prototyping
	Simulation

an emerging reality to cocreate solutions by, for example, using prototyping techniques (Whyte *et al.*, 2008; Hales and Tidd, 2009).

A central design thinking premise is that people buy objects not only to meet their needs but also because of psychological and cultural factors at work. This suggests that we need to inquire about the "meaning" of products in people's lives — and then develop ways of bringing this into the innovation process. For example, Apple's iPhone changed the meaning of a phone from a communications device to the core of a highly interactive social system, while Nintendo's Wii changed the meaning of computer gaming from a largely solitary activity to an interactive family pursuit. This is the role of design: using tools and skills to articulate and create meaning in products, as well as increase services. From this perspective, products may be new not only in terms of their technology but also in terms of their design. Combining technology-push and design-driven innovation allows for opening up a new dimension for creating radically new concepts which have meaning in people's lives (Figure 47).

Design's increasing importance as a source of innovation also engages the world of services. Joseph Pine used the term "experience economy" to describe the evolution of innovation from meeting needs toward creating experiences (Pine and Gilmore, 1999). In an increasingly competitive

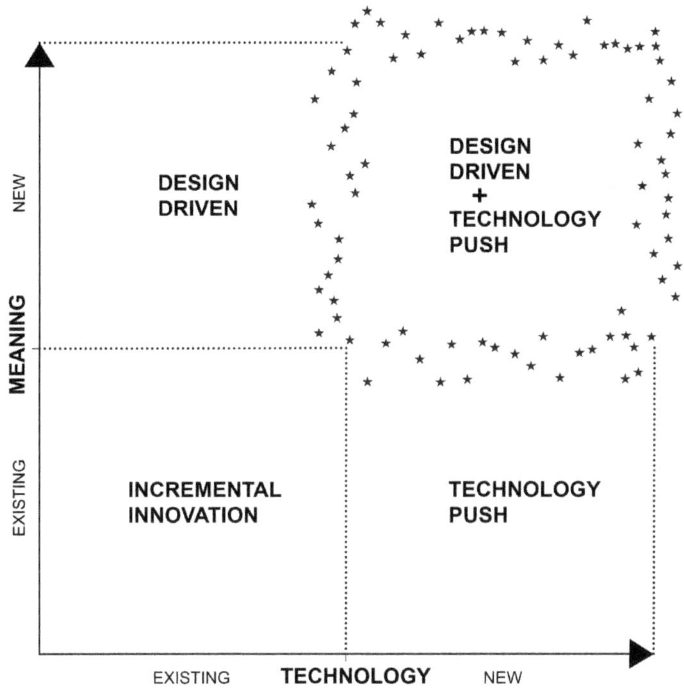

Figure 47: The role of design driven innovation in NPD.

world, differentiation increasingly originates from "experience innovation" — especially in services where fulfilling needs takes second place to the experience's meaning and psychological importance. For example, the restaurant business is moving from an emphasis on food as an essential human need to increasingly significant experience innovation focused on restaurants as systems of consumption involving the product, its delivery, the physical and cultural context, etc. Service providers, such as airlines, hotels, and entertainment businesses, are increasingly differentiating themselves along such experience innovation lines.

There is a growing emphasis on ethnography and studying what users actually do in a context, instead of what they say they do. The examples reported above represent experiments with an alternative approach to service innovation that moves away from linear supplier or consumer models to those with a "cocreation" emphasis. This approach suggests a trajectory for further development and for testing design methods' role, as well as for research into service innovation, especially in the public sector. It is important to note that this approach requires different methods and tools, such as the extensive use of scenarios and storytelling or narrative enquiry, which are better geared to capturing and exploring the system-level perspective where users and providers interact in a complex way.

Deep diving and probe and learn are two methods that help identify what potential users actually do in practice.

Customer integration in action: IDEO

IDEO, one of the most successful design consultancies in the world, is based in Palo Alto, California and in London, UK. It helps large, worldwide consumer and industrial companies design and develop innovative new products and services. Behind its rather typical Californian wackiness is a tried and tested process for successful design and development:

1. Understand the market, client, and technology.
2. Observe users and potential users in real-life situations.
3. Visualize new concepts and those customers who might use them by using prototyping, models, and simulations.
4. Evaluate and refine the prototypes in a series of quick iterations.
5. Implement the new concept for commercialization.

The first critical step is achieved through close observation of potential users in context. Tom Kelly of IDEO argues, "We're not big fans of focus groups. We don't much care for traditional market research either. We go to the source, not the 'experts' inside a (client) company but the actual people who use the product or something similar to what we're hoping to create… we believe you have to go beyond putting yourself in your customers' shoes. Indeed, we believe it's not even enough to ask people what they think about a product or idea…customers may lack the vocabulary or the palate to explain what's wrong, and especially what's missing."

The next step is to develop prototypes to help evaluate and refine the ideas captured from users. "An iterative approach to problems is one of the foundations of our culture of prototyping…you can prototype just about anything — a new product or service, or a special promotion. What counts is moving the ball forward, achieving some part of your goal."

Source: Kelly (2002).

In deep diving, market research uses innovative new possibilities. Most market research has become adept at hearing the "voice of the customer" via interviews, focus groups, panels, etc. But what people say and what they actually do sometimes differs. In recent years, there has been an upsurge in the use of anthropological-style techniques to get closer to what people need or want in the context in which they operate. "Deep diving" is one of many terms used to describe the approach; "empathic design" and "ethnographic methods" are others.

Much of the research toolkit in this context originates from the field of anthropology, where the researcher primarily aims to gain insights through observation and immersing herself in the day-to-day life of the object under study rather than through questioning. For example, to ensure their new terminal at Heathrow would address user needs well into the future, BAA commissioned research into what its users in 2020 might look like and what their needs might be. The ageing population was mentioned as an issue; by focusing on the behavior of older people at the airport, the researchers noticed that they tend to go to the toilet rather frequently. The conclusion was that more toilets should be planned for Terminal 5. However, when a researcher actually followed people around, they noted that many people going to the restrooms did not actually go to the toilet but went there because it was quiet and they could hear the announcements!

In probe and learn, market research builds on quick and iterative learning cycles. One of the problems with a radically different future is that it is hard to imagine and predict how things will unfold. Trying something out (or probing) and learning from the results, even if these do represent a presumed failure, is sometimes a powerful approach. In this way, emergent trends, potential designs, etc. can be explored and refined in a constant learning process.

There are two complementary dimensions here: first, the prototyping concept as a means of learning and refining an idea; second, the pilot scale-testing concept before this idea is moved to a mainstream market. In both cases, the underlying theme is essentially one of "learning as you go," trying things out, making mistakes but using the experience to get closer to what is needed and will work. Geoff Penney, the Chief Information Officer of the US-based investment house Charles Schwab, once said, "To avoid running too much risk we run pilots, and everyone knows it is 'just' a pilot and is not afraid of making suggestions for improvement — or killing it."

Not surprisingly, prototyping is particularly relevant in product-based firms. For example, Bang & Olufsen has revitalized its prototyping department and made it refer directly to the company's innovation hub. The prototyping department is engaged in new ideas as early as possible, and the experiences are that this strongly supports the process. After a period with disappointing results of applying electronics in toys, LEGO made a change in its development approach toward more intensive use of prototypes. Prototypes were created within days — often within hours — after the ideas matured. The result was a much more precise dialogue within the organization and with the main customers. This eventually led to more simple technology — and more success in terms of sales.

Design space's two dimensions also need to be understood. One way of looking at the design challenge is in terms of the ways in which an organization frames the innovation space. Just as human beings need to develop cognitive schemas to simplify the "blooming, buzzing confusion," from the myriad stimuli in their environment, so too do organizations make use of simplifying frames. They "look" at the environment and take note of elements which they consider relevant — threats to watch out for, opportunities to take advantage of, competitors and collaborators, etc. The construction of such frames helps give the organization stability and — amongst other things — defines the space within which it will search for innovation possibility. While there is scope for organizations to develop their individual ways of seeing the world — their business models — in practice, the commonality is often within a sector. Consequently, most firms in a particular field will adopt similar ways of framing — assuming certain "rules of the game," following the same trajectories.

These frames correspond to accepted "architectures" — the ways in which players see the configuration within which they innovate. The dominant architecture emerges over time, but once established, it becomes the "box" within which further innovation takes place. We are reminded of the difficulties with thinking and working outside this box, because the structures, processes, and toolkit — the core routines — which the organization (and its key reference points in a wider network of competitors, customers, and suppliers) has learned and embedded, reinforce the box.

This perspective highlights the challenge of moving between knowledge sets. Firms can be radical innovators, but developments outside their search trajectory can still "upstage" them. The problem is that search behavior is essentially bounded exploration, which raises two important challenges:

First, when there is a shift to a new mindset — or cognitive frame — established players may have problems, because this requires the reorganization of their thinking. It does not simply mean adding new information but changing the structure of the frame through which they see and interpret that information. They need to "think outside the box" within which their bounded exploration takes place, which is difficult, because the box is highly structured and reinforced.

Second, the new frame may not necessarily involve a radical change in technology, or markets, but instead a rearrangement of the existing elements. Low-cost airlines, for example, did not involve major technological shifts in aircraft, or airport technology, but focused on problem solving, which made flying available to an underserved market segment. Similarly, the "bottom of the pyramid" development is not about radical new technologies but about applying existing concepts to underserved markets with different characteristics and challenges. Incremental innovation may be required to make the new configuration work, but this is not usually new to the world, merely being problem solving.

A map of innovation search space helps us understand the value of design thinking and structures the different approaches to design thinking by organizing and managing the search process (Figure 48). The vertical axis refers to the familiar "incremental/radical" dimension in innovation, whilst the horizontal relates to environmental complexity or the number of elements and their potential interactions. Increasing complexity means that it becomes increasingly difficult to predict a particular state due to the elements' increasing number of potential configurations. Moving to the right introduces the problem of new and emergent architectures arising from complex elements' alternative ways of framing.

Using this idea of different frames, we can explore four zones in Figure 48. While those approaches for dealing with the left-hand side — zones 1 and 2 — are well developed, we argue that there is still much to learn about the right-hand side challenges and the methods and tools one could use to approach them in practical terms.

Zone 3 is essentially associated with reframing. It involves searching a space where alternative architectures are generated, exploring different permutations and combinations of elements in the environment. It is important to note that this often happens by working with elements in the environment that established business models do not embrace, for

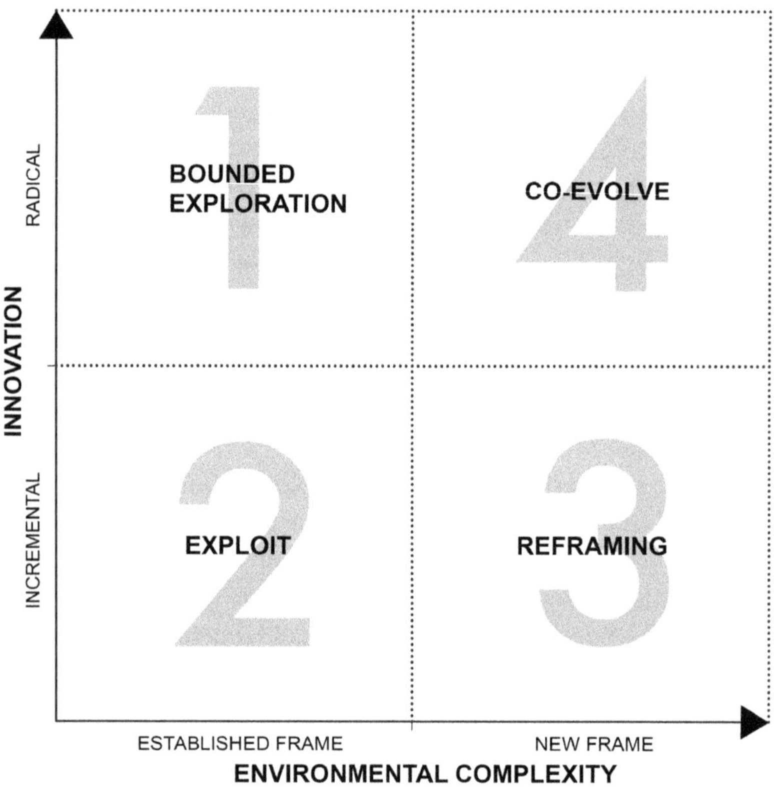

Figure 48: Innovation search space map to structure design-thinking approaches. (Reproduced with permission from Tidd and Bessant (in press).)

example, Christensen's work on fringe markets, Prahalad's bottom of the pyramid, and von Hippel's extreme users.

An example of a Zone 3 industry is the low-cost airline industry, which did not comprise the development of new product or process, because it still involves airports, aircrafts, etc. The re-framing involved identifying new customer segments, so-called "blue sky" spaces, which may have avoided flying due to the high costs, such as students and the retired, and developing an innovation system to serve these new target markets. Rethinking the business model required extensive product and process innovation to realize it, for example, online booking, fast turnaround times at airports, multiskilling of staff, etc., but the end result was a reframing and the creation of a new innovation space.

Zone 4 represents the "edge of chaos" complex environment where innovation emerges as a product of a coevolution process. This is not as much the product of a predefined trajectory as the result of complex interactions between many independent elements. Amplification and feedback processes reinforce what begin as small shifts in a direction and gradually define a trajectory. This is the pattern — the "fluid state" — before a dominant design emerges and sets the standard. Search strategies are difficult in this respect, because it is impossible to predict what is going to be important, where the initial emergence will start, and where the feedback and amplification will happen. The best an organization can do is to try and place itself within that part of its environment where something might emerge and then develop fast reactions to weak signals. In this regard, strategy can be distilled to three elements: be in there, be in there early, and be in there actively (i.e., in a position to be part of the feedback and amplification mechanisms). Consequently, very high levels of experimentation characterize this zone (Tidd and Bessant, in press).

Customer integration in action: Red Button design

Red Button Design's breakthrough product, the Midomo Water Purifier, is representative of user-led, bottom-up innovation appropriately marketed through nonlinear methods and a formula for proprofit social

(Continued)

(*Continued*)

entrepreneurship that may lead the way in our understanding of how we implement humanitarian innovation in the future. Red Button Design's code of practice, which is called "Design against Dependency," is

- designed specifically for the humanitarian sector, rather than the common practice of adapting technology from medical or military innovation
- able to address multiple related problems in a single product purchase that maximizes value
- designed to be easier to use and maintain than existing products
- environmentally preferable to existing solutions
- deployed using a business model designed for NGOs: It keeps donor funds in the community for longer, thus maximizing the "benefit for spend," and promoting community independence
- designed for local assembly, thus allowing a regional variation in product design which maximizes Midomo's utility and acceptance by users, while also helping NGOs achieve their targets of promoting local employment and economic growth.

The product is a Reverse Osmosis Sanitation System, a three-in-one water transportation, storage, and filtration system. Powered by the rotation of the wheels on which it is pushed, the Midomo purifies water by forcing it through pores small enough to only allow water molecules through. Users can visit the nearest water source, rather than walk the 5–10 km to avoid contaminated open water sources. From there, the user can easily transport the water they have collected (up to 50 l), rather than struggling with far less uncontaminated water. By the time they get back to their homes, the motion of the transportation will have purified the water. The Midomo puts the end user in a position of control; each Midomo filter should last at least a year, and if it were to stop working, the Midomo will dye the water to warn the user not to drink it.

5.2 Outcome-driven innovation (ODI)

ODI is a systematic NPD process starting with an innovation strategy and culminating in a breakthrough concept. The process takes the innovating company step by step from developing a strategy to gathering qualitative and quantitative customer insights in order to identify specific customer needs. ODI also includes suggestions to reposition the current offerings and develop new ones in accordance with the identified needs. Figure 49 depicts this process's most important steps.

In the first step, the innovating company has to decide whether it plans to address existing or new customers, and whether it wants to improve the current offering or develop a new one.

In the second step, the innovating company captures qualitative customer input. The way this input is gathered is very distinctive and forms one of the core elements differentiating ODI from other NPD processes. Ulwick (2005) criticizes traditional customer insight tools for delivering either vague needs information (e.g., improve quality) or solution information (e.g., provide a razor with a wider handle). While vague information lacks clear indicators to develop new concepts and to measure their performance relative to these needs, solution information from customers

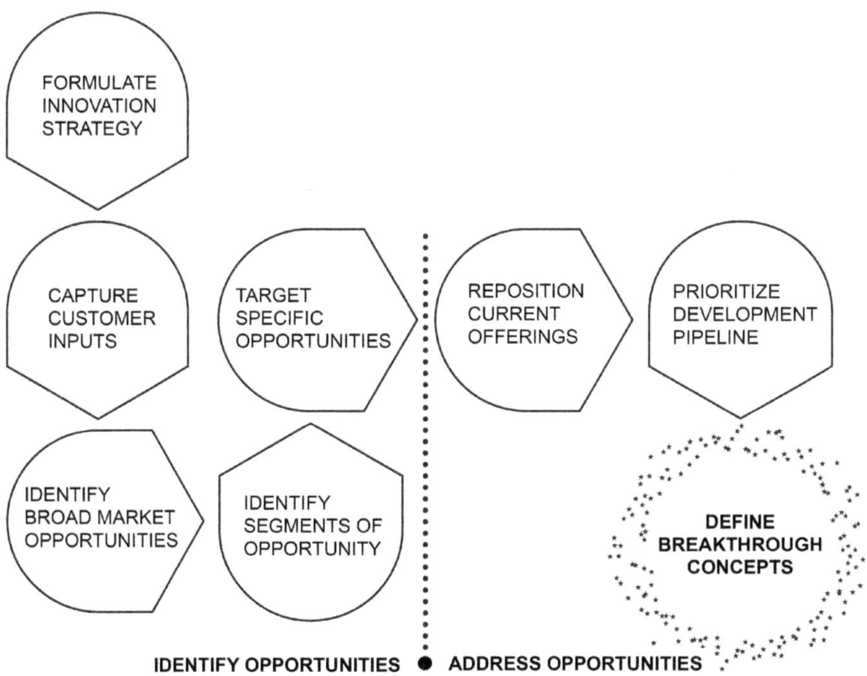

Figure 49: Steps of the ODI process.

only provide one possible solution for a problem, while alternative solutions could solve the problem in superior ways. For example, customers might have requested a wider handle to avoid the razor slipping out of their hands when shaving, although an alternative concept that would also fulfil this need might be a razor with a normal-sized, but rubberized, handle.

The outcome-driven approach strives to move beyond traditional insights by encouraging customers to frame their needs in terms of the jobs to be done. A job to be done consists of three elements: a direction (e.g., "minimize…"), a unit of measure (e.g., "…the number of times…"), and an outcome (e.g., "…the razor slips out of my hand when I shave.").

In the third step, the innovating company identifies broad market opportunities by carrying out a quantitative market research in which the respondents are given a list of the jobs to be done as extracted from step 2. On a scale (e.g., ranging from 5 (=very important) to 1 (=not at all important)), the respondents indicate how important a particular job to be done is and how satisfied they are with existing market offerings in respect to their ability to fulfill the job to be done (e.g., ranging from 5 (=very satisfied) to 1 (=not at all satisfied)). By means of an analysis, the innovating company calculates the percentage of respondents providing a 4 or 5 ranking and notes this on a scale ranging from 10 (for 100%) to 0 (for 0%). In other words, if 95% of the respondents indicated the importance rating of a particular job to be done by ticking 4 or 5, this would result in the entry of 9.5 on this scale. Based on these scales of the importance and satisfaction ratings, we calculate the following opportunity algorithm:

$$\text{Importance} + \max(\text{Importance} - \text{Satisfaction}, 0) = \text{Opportunity}.$$

As illustrated in Table 27, applying this algorithm leads to an opportunity score. Scores above 10 represent underserved outcomes that are attractive innovating opportunities.

Scores below 10 are either overserved, or appropriately served, outcomes on which the company should not focus its innovation endeavor, as they do not serve as unique selling propositions. More precisely, Figure 50 shows that outcomes with a higher satisfaction than importance score are overserved outcomes, while appropriately served ones are either outcome combinations with low importance and satisfaction ratings or

Table 27: Calculating the opportunity algorithm.

Outcomes	Importance	Satisfaction	Applied algorithm	Opportunity score
Desired outcome 1	8.20	4.50	8.2 + (8.2 – 4.50)	11.90
Desired outcome 2	7.30	9.10	7.3 + (0)	7.30

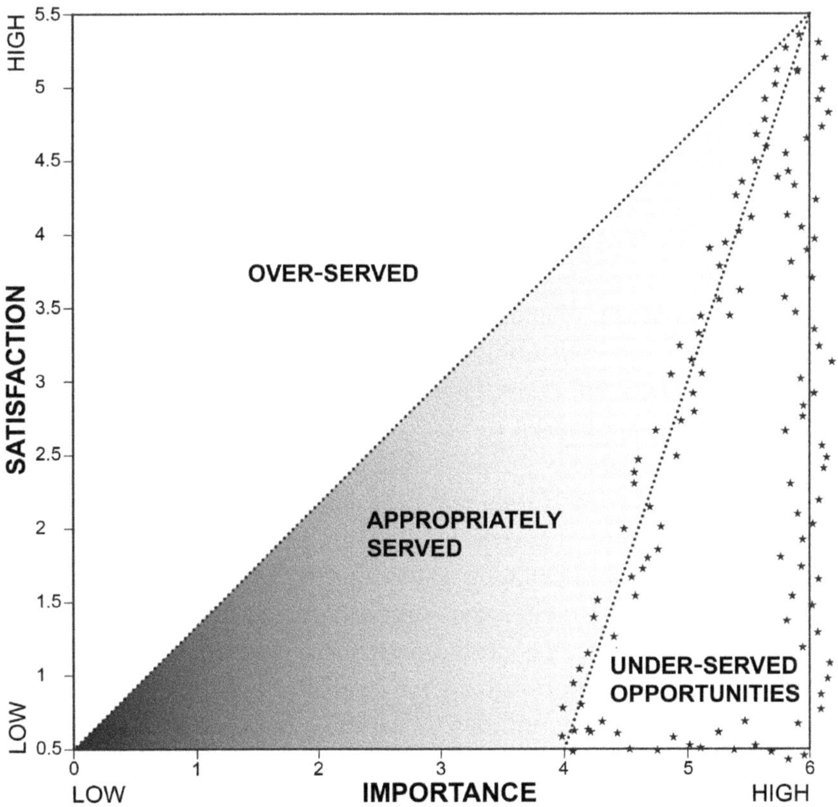

Figure 50: Visual representation of how to identify opportunities based on importance and satisfaction ratings. (Adapted with permission from Ulwick (2005). Copyright © 2005 The McGraw-Hill Companies, Inc.)

high importance and satisfaction rating. Attractive opportunities are outcomes that are highly important to customers, and customers are dissatisfied with the way the current offerings address these outcomes.

Table 28: Strengths and challenges of ODI.

Strengths	Challenges
• Reduces the risk of the NIH syndrome, as customer input is presented in a way that encourages NPD teams to use their creativity to solve jobs to be done • Qualitative and quantitative customer insight is gathered • Taking the importance and satisfaction scores of jobs to be done into consideration ensures focus on features that are highly relevant for customers	• Formulating needs in the jobs to be done format can be challenging for customers • Jobs to be done are difficult to formulate in terms of emotional elements and hedonic products

In steps 4 and 5, the innovating company divides the market into segments based on the outcome different customers strive to achieve and target the most attractive market segments. Segments are groups of customers with unique sets of underserved or overserved outcomes. Customers within a segment are both similar with customers within the segment and different from those of the other segments. The segments with underserved outcomes that can be served with an attractive revenue potential form the target market.

In steps 6–8, the innovating company addresses the identified opportunities by focusing on its current offerings and developing new offerings. This endeavor includes designing a suitable communication strategy for a specific outcome, evaluating the current product innovation pipeline according to the product's contribution to satisfying the identified attractive opportunities, and reprioritizing the products in the pipeline according to their fit with these opportunities. In addition, these steps include generating new product ideas catering to the identified opportunities and developing them into concepts.

Table 28 highlights the main strengths and weaknesses of ODI.

5.3 Quality function deployment (QFD)

QFD is a method for planning and developing a product's features based on customer requirements that originated in Japan in the sixties.

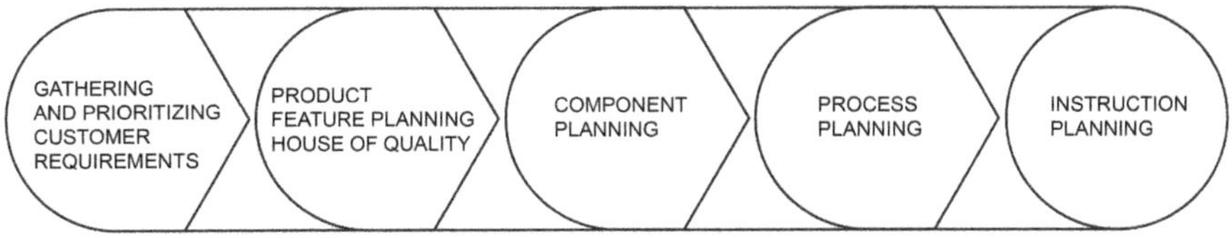

Figure 51: The five QFD phases.

QFD strives, by means of a rigorous process, to transform the voice of the customer into the voice of the engineer. In QFD thinking, transformation matrices systematically change customer requirements into product features for product planning, product component manufacturing, and production and process planning (Akao, 2004).

QFD processes follow five phases, as shown in Figure 51, with the output of each phase serving as input for the next phase. In the first phase, the innovating company surveys customers to identify their requirements. The QFD method does not prescribe a specific way of gathering customer information. Information can come from systematic interviews, unsystematic customer interactions at fairs, an analysis of customer complaints data, or from methods to identify latent needs. The quality of the data that enter the later stages of the QFD process can thus vary largely. The requirements also need prioritization, which is best decided by asking customers directly to indicate the importance of certain features. Alternatively, practitioners sometimes prioritize requirements in a company-internal joint workshop of customer insights specialists and technical experts.

Marketing and R&D employees usually need to discuss and interpret the gathered information before the requirements and their prioritization can function as the basis for the next phase, during which the innovating company translates customer requirements into product features. This translation takes place in the "house of quality." This "house" is the centerpiece of the QFD process with the translation requiring 10 steps.

The house describes the customer requirements on the horizontal axis and the specific product features on the vertical axis. The intention of this breakdown is to integrate the market perspective and the technology perspective. Figure 52 shows the basic outline of the house of quality. Figure 53 is a typical representation of a house of quality, with the numbers referring to the 10 steps necessary to build it. Step 1 constitutes the aggregation and prioritization of the customer choice. Step 2 provides data on the company's existing products as compared to the competitors' products based on the customers' perceptions of their strengths and weaknesses. Step 3 leaves room for notes on the service and the distribution. Step 4 involves the translation of requirements into features. The innovating company has to describe the product features' desired target values in a measurable way and with a clear indication whether fulfilling the required feature quality requires a minimizing of, a maximizing of, or retaining the current quality level. For example, the loading process is a product feature of a battery charging system and the desired target value could be a lifetime of 20,000 h.

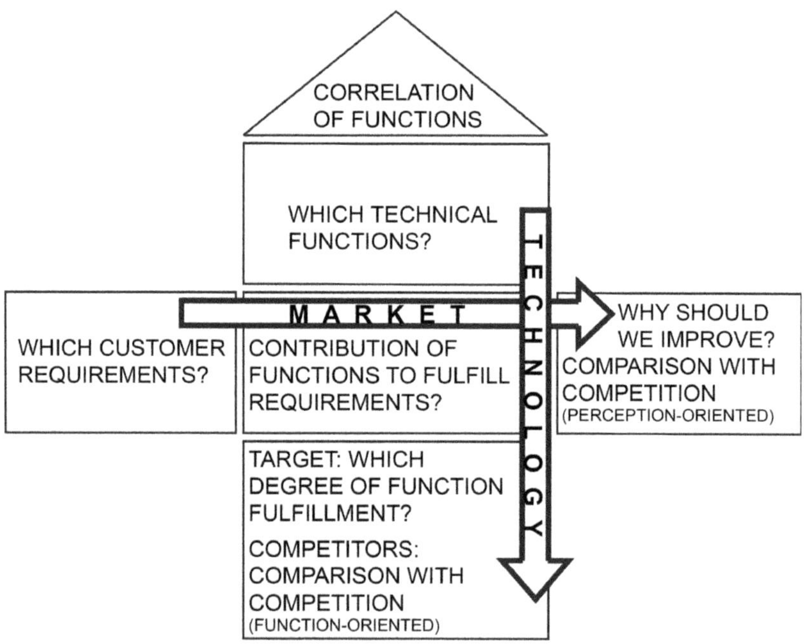

Figure 52: Outline of a house of quality.

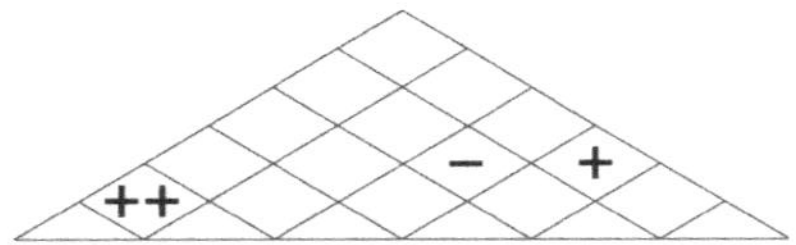

ENGINEERING CHARACTERISTICS

DIRECTION OF OPTIMIZATION

CUSTOMER REQUIREMENTS	PRIORITY	SIMPLE EXCHANGE OF COMPONENTS	MODULARITY	CASING SIZE	CHARGING PROCESS	SAFETY FUNCTION	AUTOMATIC GATHERING AND ANYLSIS OF INDICATORS	SUM OF ROWS	COMPETITIVE EVALUATION: CURRENT OFFERING	COMPETITIVE EVALUATION: COMPETITOR A'S OFFERING	COMPETITIVE EVALUATION: COMPETITOR B'S OFFERING	OTHER DETAILS
DIRECTION OF OPTIMIZATION		▲	▲	▶	▲	▲	▲					
LONGER LIFETIME OF CHARGING SYSTEM	8.0		3	3	1	9		128	3	4	1	NOTE `A`
HIGHER LIFETIME OF BATTERY	6.8				9	3	1	88.4	2	3	3	
SIMPLE SETTING UP	3.3			9	1	1		36.3	4	2	5	
FASTER SERVICE	6.3	9	9	9				170.1	3	3	3	NOTE `B`
INVESTMENT SAFETY (TEC. SUPPORT, PARTS)	5.5	3	3	1		3	3	71.5	4	2	3	
TECHNICAL IMPORTANCE (ABSOLUTE)		73.2	97.2	115.9	89	95.7	23.3					
TECHNICAL IMPORTANCE (RELATIVE)		14.8	19.7	23.4	18	19.4	4.7					
TARGET VALUE/SPECIFIC VALUE (ILLUSTRATIVE EXAMPLE / HOURS)						20,000						
COMPETITIVE EVALUATION (1 WORSE, 3 EQUAL, 5 BETTER THAN COMPETITORS)		3	5	5	1	5	3					
TECHNICAL DIFFICULTY (1= LOW; 5 = HIGH)		2.5	2	2.5	4	1	5					
OTHER NOTES			NOTE `C`			NOTE `D`						
PRIORITY AFTER EVALUATION		2	1	1	2	1	3					

Figure 53: Example of a house of quality for battery charging systems.

In step 5, the relationship between the customer requirements and the product features is established. In this matrix, an innovating company indicates whether the realization of a certain product feature contributes weakly or strongly to the fulfillment of a certain customer requirement. The relation is generally shown as a number between 1 and 9, with 1 indicating a weak relationship and 9 a strong one. Step 5 also requires the multiplication of each column's values with the weights of the customer requirements' priority and their aggregation to calculate the product features' absolute and relative relevance. In step 6, the top of the house of quality is scrutinized. The top is the abstract representation of a rooftop and contains indicators of the interplay of the product features: "+" or "++" signs respectively point out a positive and a strongly positive relationship between two features, while "–"/"––" signs respectively point out a negative and a strongly negative one (Daetz *et al.*, 1995; Walker, 2002). In the example in Figure 53, the loading process correlates positively with the automatic gathering and analysis of indicators.

Step 7 examines the bottom of the house to compare the company's current product features with those of competitors. In step 8, the innovating company indicates potential challenges in realizing the target values of each product attribute. Step 9 leaves space for general notes worthy of inclusion in the overall evaluation of the product attributes. In step 10 the innovating company finalizes phase two by checking the results, evaluating the product features, and selecting the highest rated product features for further development. To prioritize product features, the company can draw a portfolio (see Figure 54) with the relative importance and the difficulty of the technical realization as axes and then split the matrix in the middle, thus quartering the matrix. The first priority is attributed to features that fall within the quarter with a high technical feasibility and a high relative importance. Features have second priority if they are either in the high relative importance/low technical feasibility quarter or in the low relative importance/high technical feasibility quarter. The third priority features would then be those in the low relative importance/low technical feasibility quarter. The portfolio should serve as the basis for the QFD team members' discussion of the

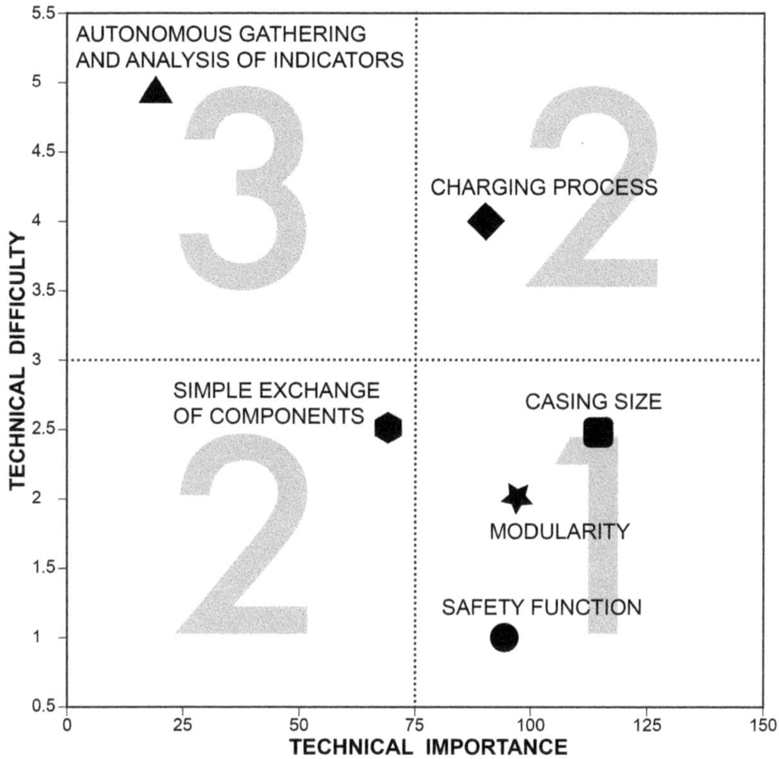

Figure 54: Portfolio to facilitate product feature prioritization in the house of quality.

product priorities, rather than be regarded as an inflexible measure of priority (Prielinger, 2011).

For accurate results, phases 1 and 2 require intensive interaction at the R&D-marketing interface to translate (blurred) customer requirements into quantifiable product features. The planning of phases 3–5 is less resource consuming than the first two. In the third phase, the innovating company plans the components required to realize the selected features. The relevant product features, which were compiled in the QFD house and provide insights into what to produce, are correlated with the requirements for components and modules that clarify how to produce these features. In the fourth phase, the innovating company compares the components' and modules' requirements with the production process's characteristics to clarify how these requirements will be met. In the fifth phase, the innovating company uses a matrix to compare the production processes, as well as the work and testing instructions, to ensure that the instructions meet the production processes' requirements.

Customer integration in action: Transforming customer requirements into energy for next-generation charging systems

Fronius International GmbH is a premium supplier of technologies for monitoring and controlling energy for welding machinery, solar converters, and battery charging systems. In the latter area, it offers solutions for traction (used by intralogistics in electric floor-borne vehicles) and starter batteries (used in cars). Not only battery manufacturers but also manufacturers of forklifts and similar vehicles are immediate customers for traction batteries. End users are companies using 10 or more electric floor-borne vehicles for intralogistics. When developing a new generation of traction battery charging systems, Fronius applied QFD with the goal to develop features that were a clear unique selling proposition and would simultaneously decrease the charging systems' production costs by 10%. In a first step, Fronius gathered its immediate customers' and end users' requirements. Figure 53 provides an excerpt of the requirements and their transformation into product features in the house of quality.

The realization of the QFD was challenging, because it required several QFD team workshop meetings, which took place over a period of several weeks instead of days, due to the team members' busy agenda. This required time-consuming updates at the beginning of each workshop. Unclear formulations of the requirements and the differing degrees of their abstraction levels led to time-consuming discussions that were sometimes enlightening and sometimes unproductive. Overall, the QFD improved the information exchange between the departments, the R&D employees' understanding of the customer, and the interfunctional information sharing of the expected future product features, which allowed the employees involved in later NPD stages to be prepared in advance. In turn, the QFD application helped Fronius reach the development goals of the new generation charging systems (Prielinger, 2011).

While more popular for incremental improvements of technically complex physical products, QFD also has the potential to be used in services and in offerings of any type of complexity and innovativeness. Its utility depends largely on the customer requirements' selected and weighted quality and the translation into product features that is carried out. It is worth noting here that the translation into product features is quite straightforward in the case of incremental product improvements, because the required features can be easily understood from prior product versions and competitive offerings. These translations can be quite challenging regarding more radical innovations and might require additional methods.

This holistic approach allows for the use of clear and measurable target values for different features and makes the positive and the negative relationships between the features transparent. This transparency facilitates the allocation of limited R&D resources to the realization of the prioritized product features. It focuses R&D resources on those features that are important for the customer and avoids the overengineering of features with little customer value. As a tool to accompany the NPD planning process, QFD brings consequent and continuous customer orientation into the NPD process and also facilitates communication between R&D, marketing, and production. Consequently, it can reduce the number of change requests during the development, reduce the development time, shorten the time to market, and improve the innovation's market fit.

QFD also suffers from downsides. First, the holistic realization of all four phases is time consuming and requires method experience and considerable human resources. Second, the systematic procedure curtails the R&D team's ability to use more intuitive and creative ways of finding solutions to customer requirements. Third, overall quality is strongly dependent on the quality of the execution of elements (e.g., the quality of the formulation of the customer requirements, the quality of the translation of the requirements into features), with QFD only providing little guidance for their accurate execution (Xie *et al.*, 2003; Akao, 2004). Table 29 summarizes QFD's key pros and cons.

Table 29: Strengths and challenges of QFD.

Strengths	Challenges
• Entices consequent and continuous customer orientation throughout the innovation process • Improves discussions and understanding across functional boundaries • Efficiency and time-to-market gains	• Resource intensive • Risks curtailing creative freedom • Accurate execution of certain steps and phases is highly challenging

5.4 Market scrums

Market scrums combine the principles of agile project management with customer insights through user stories. Agile project management focuses primarily on managing customer needs and evolving requirements by fostering interaction between interdisciplinary team members and by using short development cycles (iterations), as well as continuous change and adaption throughout the project life cycle (Barlow *et al.*, 2011; Tidd, 2014). The scrum method is an agile method that helps teams solve complex problems, has its origins in software development, and is now applied for managing NPD and NSD, as well as for other managerial tasks (Sutherland, 2015). Market scrums are usually combined with Kanban, another agile method, whose strength lies in reducing unfinished work.

Figure 55 provides an overview of the set of particular steps that need to be performed within specific time boxes. The figure also indicates the three roles (scrum team, scrum master, product owner) participating in market scrums. *Scrum teams* consist of three to nine members from different functions and disciplines and work exclusively for the scrum project during the entire project. These teams are self-organizing and fully responsible for the tasks they carry out. In addition to the scrum team, the scrum method requires a *scrum master* and a *product owner*.

The scrum master coaches the team and makes sure that the team has the necessary conditions and resources to work properly but neither prescribes ways to solve a problem nor assigns tasks to the team members.

The product owner is part of the team, in a boundary position between the team and the external stakeholders (top management, customers), and

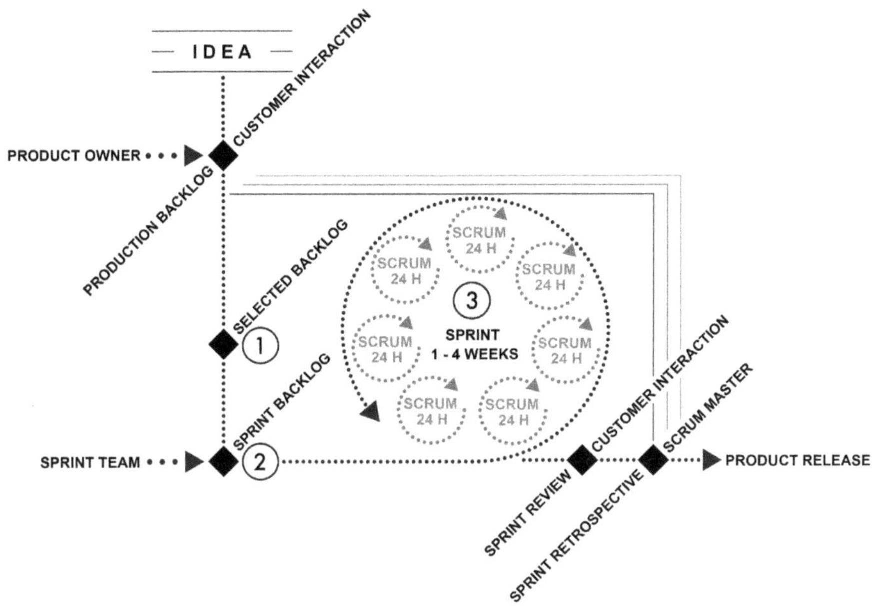

Figure 55: Steps and roles in market scrums.

is responsible for the scrum initiative. In this key role, the product owner writes and manages the product backlog, through which he or she acts as a representative of the customer's voice and ensures that the customer's interest guides the project. The product backlog includes a list of user stories (i.e., specific functionalities from a user perspective) that the product owner writes, which is based on customer interaction and team discussions. These user stories are short (one to two sentences long) descriptions of the user requirements. User stories describe the added value from a customer perspective. Each user story describes one clear and concise added value in terms of the user's role, the target, and reason why this target was chosen. A possible user story for a mobile-health application for cancer patients would be, "As a patient (user role), I want to take a photo of the needed meds, with the app automatically identifying the name and dosages of the prescriptions (target), so that I always have a list of my medications and find it easier to take the correct amount at the correct time point (reason why)." The user role is not limited to the person actually using the app, which is the patient in the cancer app case, but may also include other users, such as physicians, national health agencies, and pharmaceutical firms.

The product owner also has to prioritize user stories, dividing them into "should have" and "must have" ones. At a minimum, the final product has to include all must haves, because the new product would not function without them. The inclusion of should haves depends on the time and resources left. The product owner also arranges the user stories so that they are meaningful for the development task; if one user story (e.g., the login or logout on a webpage that connects users of the mobile-health app with physicians and with other patients) depends on another one (e.g., registering on a webpage), the latter has to appear on the product backlog list before the first one. A user-story-based backlog facilitates assessing product requirements' relevance, prioritizing the backlog, and taking decisions. The product backlog is a dynamic document; if necessary, the product owner can add, adapt, or delete user stories and change the priorities at any time throughout the project.

A scrum project starts with a clearly formulated idea for a new market offering, for example, a mobile-health software that allows patients to understand their disease better and learn about the medications' side effects that facilitates the day-to-day management of the therapy and helps physicians optimize individual medication plans based on the provided data. Based on this idea, the product owner writes a concept paper that describes the app's key idea, basic functions, and the way it generates value for the company and the user. The product owner also writes the product backlog. The product owner does this effectively by gathering customer insights and discussing user stories with the team members in a joint half-day or full-day market sprint planning workshop. The other option is a two-stage process, during which the product owner first discusses the project with the customer and then with the team. In these workshops, the participants create a mind map of the specific project, or another visualized representation of the key ideas and requirements, from different users' perspectives (e.g., physicians, patients). The discussion of these ideas and requirements by the product owner, the customer, and the team facilitates the creation of user stories, which eventually form the product backlog. Once the user stories have emerged from the discussion, the workshop participants assess the required effort for each user story.

The effort is assessed not in terms of the time needed but in terms of its relative size. Stories considered to need less time than others are short stories and receive 1 story point, large ones receive 5, and middle ones 3 points. Relative size is used, because it tends to be easier to assess than precise absolute time spans.

After the initial planning workshop, the product owner writes the product backlog and the scrum team executes its first sprint. A sprint is a development spanning a maximum period of 4 weeks. The first step of the sprint is the sprint planning meeting, which can last several hours or a day. In this meeting, the scrum team selects and plans the user stories that they will realize in this sprint. The project owner presents each user story and discusses this with the team. The discussion of the user stories allows all the team members to understand and scrutinize them. The product owner also presents the planned relative size, and the scrum members discuss and adapt this if necessary. The team members decide whether they want to include a specific user story in a particular sprint. As soon as they decide not to include the next user story on the list, the *selected backlog* is full and the team stops the discussion of the remaining user stories on the backlog. For example, a team might choose a *velocity* of 6 for their current sprint. This signifies that they will include user stories until 6 *story points* have been reached. Velocity is a self-adapting value that the team members change from one sprint to another, depending on their experience with prior sprints.

In the second part of the sprint planning workshop, the product owner writes the selected backlog's user stories on cards and the *confirmation criteria* on the reverse side. These are a set of criteria that the team needs to meet before the product owner states that a specific user story task has been realized. These criteria are listed in a precise form. An example of a criteria in the right form is "Photo of medication has to be linked to a fitting entered medication name." Normally, the product owner develops these criteria upfront through discussions with the customer and the team members. Next, the product owner clips these on a Kanban board (Figure 56), which is a whiteboard with space for the user stories and three columns: the first for the "task," the second for "in progress," and the third for "done." By dividing stories into single tasks, team members then create cards with tasks that they need to fulfil in order to

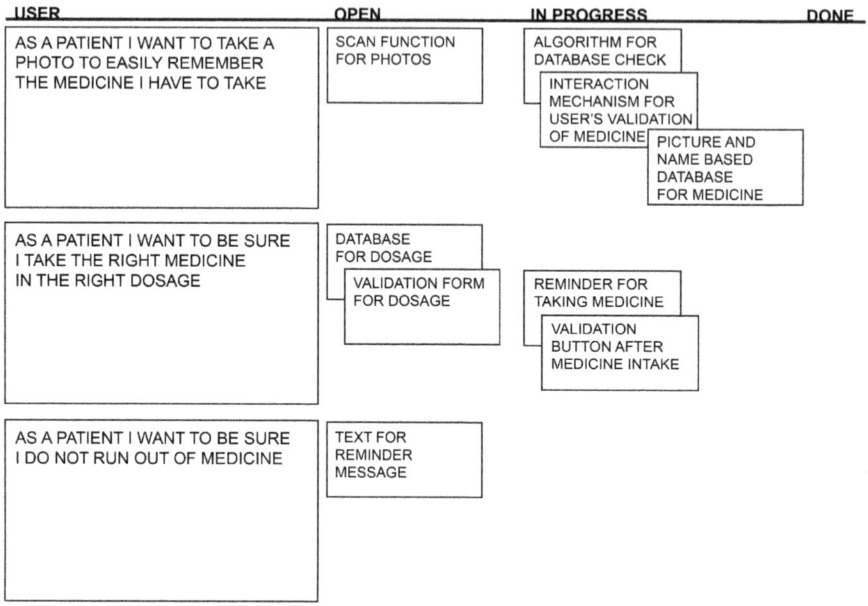

Figure 56: Example of a sprint backlog on a Kanban board.

realize the user story. The selected backlog and these tasks represent the *sprint backlog*. These task cards are first of all allocated to the task column. Individual team members choose the task on which they want to work and then add the task's card to the "in progress" column. Once finished, they allocate the card to the "done" column.

Daily scrums take place throughout a sprint's run. Scrums are standup meetings in front of the Kanban and should not exceed 15 min. In these short meetings, the team members discuss what they achieved the day before, what they plan for that day, and what impediments made their work difficult. The impediments represent tasks for the scrum master.

As soon as all the tasks of a specific user story have been completed, the functioning of the user story is tested, and the product owner approves it, after which the next one starts. At the same time, the burndown chart comes into play. As depicted in Figure 57, this chart measures and visualizes progress based on a current sprint's total story points by showing the time needed to realize a user story and by deducting the total story points

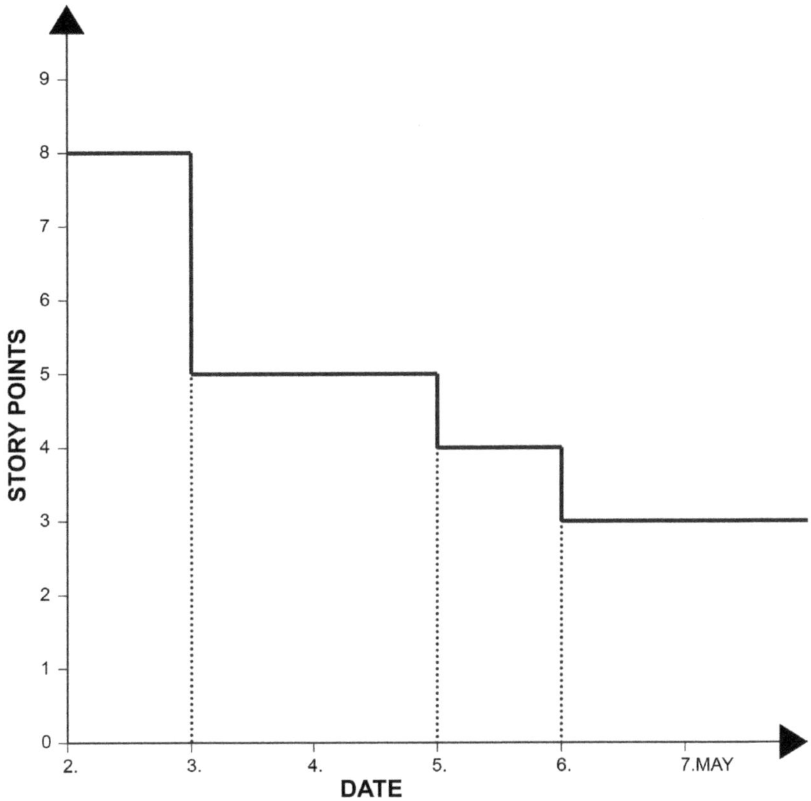

Figure 57: Example of a sprint burndown chart.

attributable to this user story from the total (Wirdemann and Mainusch, 2017; Sutherland, 2015).

When the set time for the sprint ends, the team runs a scrum review to demonstrate the realized stories' functioning. Customers may be invited to this review. Thereafter, the scrum master carries out a scrum retrospective workshop. In this workshop, the team members record the deviations between the planned and the realized story points and discuss work done well in the previous sprint and work in need of improvement. In terms of the latter, the team members make suggestions for a change for the better and then select a list of suggestions to follow in the next scrum. Thereafter the next scrum starts.

In their normal format, market scrums follow a release plan with a clear launch date. The time before this date is divided into a number of scrums. The release date is never postponed; instead, a prototype not containing the full number of planned user stories is released (Sutherland, 2015; Mullins and Sutherland, 1998).

Customer integration in action: General Electric's (GE's) market scrums

GE's Software Sciences and Analytics unit's core task is to develop information-based software systems for internal and external clients in diverse areas ranging from wind farms to aircraft engines. The GE unit's market scrums are yearlong development processes that, right from the start, involve internal or external business clients when they, together with GE's scrum team members, receive training on the unit's agile planning principles. In addition, if required, the dedicated scrum team involves in-demand experts for particular subtasks. In the first scrum, the deliverables are not pieces of prototypes but pieces of information, which are central for GE's Software Sciences and Analytics unit to ensure that the project answers the right questions.

If the project is top priority for the clients, they function as product owners. If not, an internal ambassador represents them. Whatever the case, clients are regularly and frequently involved in the project; they generally provide feedback every fortnight. Short feedback cycles ensure the project's credibility is conveyed to the clients and allow GE to iterate and adapt the project in a timely manner, instead of spending months on perfecting a prototype only to find that it does not meet a client's expectations.

Source: Haiken (2016).

Table 30 offers an overview on the key strengths and weaknesses of market scrums. A market scrum differs from traditional waterfall models of planning in as far as it does not, from the start, precisely define when and in which order which results should be delivered. It also lacks clear upfront planning of each required step. On the contrary, it only provides a vague framework of user stories. A detailed plan is developed in a dialogue-based form at the beginning of a sprint and only includes activities with which that sprint deals. This process ensures the timely production of user story details and increases the flexibility and the spontaneity

Table 30: Strengths and challenges of market scrums.

Strengths	Challenges
• Transparent, efficient processes • Self-organization in teams fosters employees' motivation and personal responsibility • Regular customer feedback ensures market fit • User stories focus NPD team on added value for customer • Allows for market learning based on early prototypes followed by improvements for revised and updated versions	• Focus on small customer base • Difficult for NPD with highly interrelated and dependent modules • Customers may use insights into the innovating company's NPD opportunistically • Favoring the timely release of early prototypes rather than launching finished ones does not work for all industries (e.g., life sciences)

of the development. In addition, unplanned user stories and tasks can be flexibly added to the sprint backlog. Market scrums thus reduce time-consuming front loading, repetitive planning, and documentation work. Since cross-functional teams work together and simultaneously, instead of successively, scrum teams are in a good position to accelerate the innovation process. A downside is that scrums can only be used if complex tasks are dividable into rather independent modules.

The scrum method can improve team productivity and employee satisfaction, because the scrum members carry out scrums democratically and in a self-organizing way; this freedom usually motivates knowledge workers. The process and the progress toward target fulfillment are transparent for all the members to see, as well as for external stakeholders, through visual progress reports and burndown charts, which add to the process's clarity and efficiency.

Constant customer feedback loops allow for focusing on tasks to which the customer adds value. Customer-based feedback loops work especially well with products like software, because beta versions can be released to a certain customer group, as well as with customized innovation projects that focus exclusively on a specific customer (see *Customer integration in action* box). Scrum projects can therefore lead to more efficient development processes, faster time to market, and an innovation's better market fit. On the other hand, customers' intense involvement may make them aware of the mistakes that the scrum team made, which they might exploit by, for example, asking the innovating company to cover costs supposedly due to such mistakes. Furthermore, market scrums only allow the close integration of a relatively small number of willing customers. This is challenging for innovations not limited to the involved customers as clients but which address a larger market. The selection of some customers in market scrums could then bias the team toward just focusing on a nonrepresentative part of the entire target market, ending up with an innovation that is only relevant to a niche market.

Customer integration in action: The role of customers in lean management for business model innovation

In his seminal book, Ries (2011) presents a lean process for business model innovation that start-ups inspired and which is based on early customer interaction: Once a vision has been developed for the business model, it needs to be translated into falsifiable hypotheses. Next, the innovating company needs to create a *minimum viable product*. This product contains the smallest number of features required to test the hypotheses. This minimum viable product is then presented to the customer. Based on the customer feedback, the hypotheses are validated or rejected. Based on these insights, the product is either developed further and scaled up, scrapped, or adapted. At the core of this approach, the innovating company does not solve disagreements through debates but by means of early and short experiments with preliminary prototypes. Prototypes are launched immediately if customers are excited about the prototypes in the experiments. This process works well in the software industry, where the prototype featuring only basic features can be launched and updates can continuously provide added value. Given the increasing number of hybrid and digital innovations in the more hardware-driven industry, this approach might become increasingly relevant for other industries. Nevertheless, this process has challenges: Although bugs seem to be an accepted feature of new software products, with users seeming to compete to find and report them, thus allowing the software to evolve and mature, products with bugs can backfire in other industries. Recent examples of mobiles catching fire and expensive product recalls indicate the potential negative effects of such strategies in respect of a company's reputation and bottom line.

Source: Ries (2011).

5.5 R&D alliances and NPD networks with customers

Organizations collaborate for many reasons: to reduce the cost, time, or risk of access to new and unfamiliar technologies or markets. However, the partners' motives and preferences will determine the precise form of the collaboration, although the nature of the technologies and markets will constrain their choice, specifically the degree of knowledge complexity and tacitness, as well as how widely available such knowledge is (Figure 58).

5.5.1 *Contract*

Resource dependence (Pfeffer and Novak, 1978; Hallen *et al.*, 2014) and agency theory (Hill and Jones, 1992; Reuer and Ragozzino, 2006) are

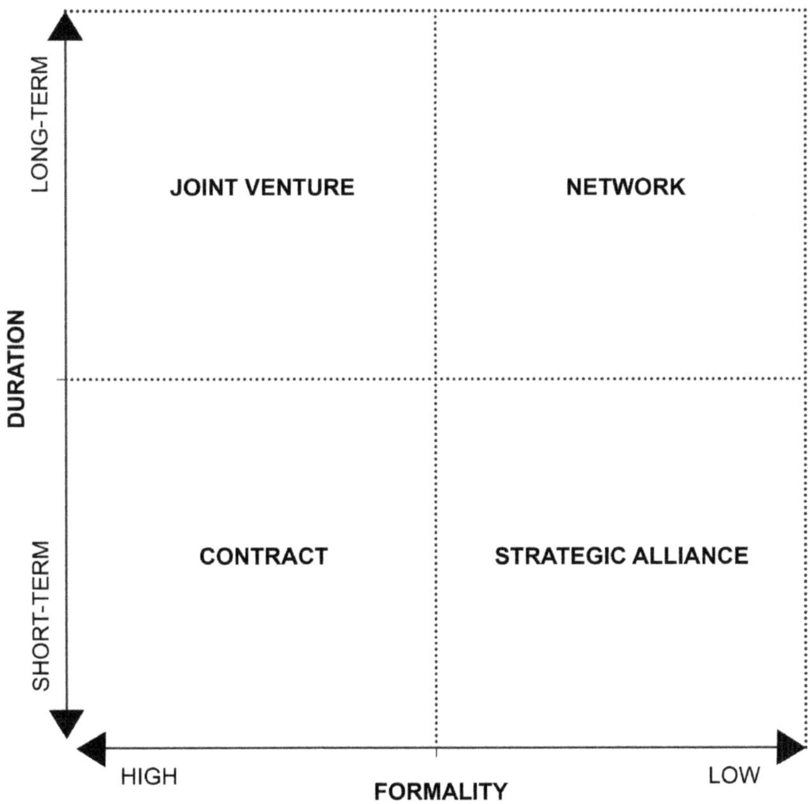

Figure 58: Formality and duration of collaboration.

more commonly used to explain vertical relationships and are concerned with the need to control key technologies in the value chain. Perceptions of Japanese manufacturers' practices have led many firms to form closer relationships with suppliers. Closer links between firms, their suppliers, and customers may indeed help reduce components' costs through specialization and the sharing of information on costs. However, factors such as the selection of suppliers and users, the timing and mode of their involvement, and the novelty and complexity of the system being developed may reduce, or negate, the benefit of close supplier–user links.

The quality of the relationship with suppliers and the timing of their involvement in development are critical factors. Traditionally, such relationships have been short term, contractual arm's-length agreements focusing on costs, with little supplier input into the design or the engineering. In contrast, the "Japanese" or "partnership" model is based on long-term relationships, and suppliers make a significant contribution to the development of new products. The latter approach increases the visibility of the cost–performance trade-offs, reduces the time to market, and improves the integration of component technologies (see the case of TSMC in the following box). In certain sectors, particularly machine tools and scientific equipment, there is a long tradition of collaboration between manufacturers and lead users in the development of new products.

Customer integration in action: Taiwan Semiconductor Manufacturing Company (TSMC)

TSMC was established in Taiwan in 1987, becoming the world's first dedicated semiconductor foundry (commonly called a fab). This "pure-play" foundry business represented a novel business model, because, unlike conventional vertically integrated manufacturers, TSMC's customers are fabless semiconductor design houses such as Qualcomm, Broadcom, and NVIDIA. Some conventional fab companies, like Intel, also outsource some of their production to TSMC. The cost of building and operating fabrication facilities has become prohibitive for all but the very largest companies, such as Intel and Samsung, especially in the case of complex logic applications. Even Advanced Micro Devices (AMD) separated its design and manufacturing businesses in 2008.

(*Continued*)

(Continued)

TSMC's headquarters and main fab plants are located in Hsinchu, Taiwan, but it also operates two wholly owned subsidiaries, WaferTech in the USA, TSMC China Company Limited, and a joint fab venture in Singapore called SSMC. Its core business is mask production, wafer manufacturing, assembly, and testing, but it also provides design and prototyping services. In 2010, TSMC joined the top 10 of semiconductor R&D spenders by spending USD 945 million, which is equivalent to 7% of its sales (called the R&D intensity), the highest of any pure foundry business. By comparison, the number one R&D spender in the industry that year was Intel with USD 6.6 billion (17% of sales), with Samsung in second place with USD 2.6 billion (8% of sales).

In 2011, the company's production capacity reached 13.2 million of eight-inch equivalent wafers, and it had more than 450 customers manufacturing more than 8,300 products for computer, communications, and consumer electronics applications. In 2012, a partnership between TSMC and Apple began production of the A5 (dual core) and A6 chip for Apple's next-generation iPads and iPhones. TSMC has benefited from the growth in smart mobile devices: It is estimated that every tablet sold globally contributes about USD 7 to its income. In 2011–2012, TSMC had sales of USD 14 billion, and by specializing in high-technology, capital-intensive contract manufacture, it maintained high gross profit margins of around 40%, although its profitability is dependent on matching its capacity and the demand closely.

In this case, some form of "partnership," or "lean" relationship, is often advocated, which is based on the quality and the development lead-time benefits experienced by Japanese manufacturers of consumer durables, specifically cars and electronics. There are a number of defining characteristics of such a partnership or "lean" supply relationship:

- fewer suppliers, longer-term relations
- greater equity — real "cost transparency"
- focus on value flows — the relationship, not the contract
- vendor assessment, plus development
- two-way or third-party assessment
- mutual learning — share experience, expertise, knowledge, and investment.

These principles are based on a distillation of the features of the best Japanese manufacturers in the automobile and electronics sectors and more recent experiments in other contexts, such as aerospace in the UK and the USA.

Other contractual bases for relationships feature intellectual property, which commonly involves licensing agreements. Licensing can have a number of benefits by allowing for the following:

- reducing or eliminating production and distribution costs and risks
- reaching a larger market
- exploiting other applications
- establishing standards
- gaining access to complementary technology
- blocking competing developments
- converting competitors into defenders.

Considerations when drafting a licensing agreement include the degree of exclusivity, the territory, the type of end use, the period of the license, and the type and the level of payments (e.g., royalty, lump sum, or cross-license). In addition, the relative investment and the weight of the share of the risk have to be calculated. Finally, the cost of licensing should be compared to alternatives to a formal legal agreement, for example, the cost of imitating the innovation or cost of litigation for infringement. There is no "best" licensing strategy, because it depends on the organization's strategy and on the nature of the technology and markets, as illustrated by Apple and Google's opposing approaches.

Research note: Does external knowledge complement or compete with internal knowledge?

Using panel data from 325 firms over 5 y, Denicolai and colleagues (2016) identified the relationship between the use of external knowledge and sales growth over time (Figure 59). The vertical axis is a measure of sales growth and the horizontal axis the proportion of external knowledge used. The research suggests that increasing external knowledge use complements

(*Continued*)

(*Continued*)

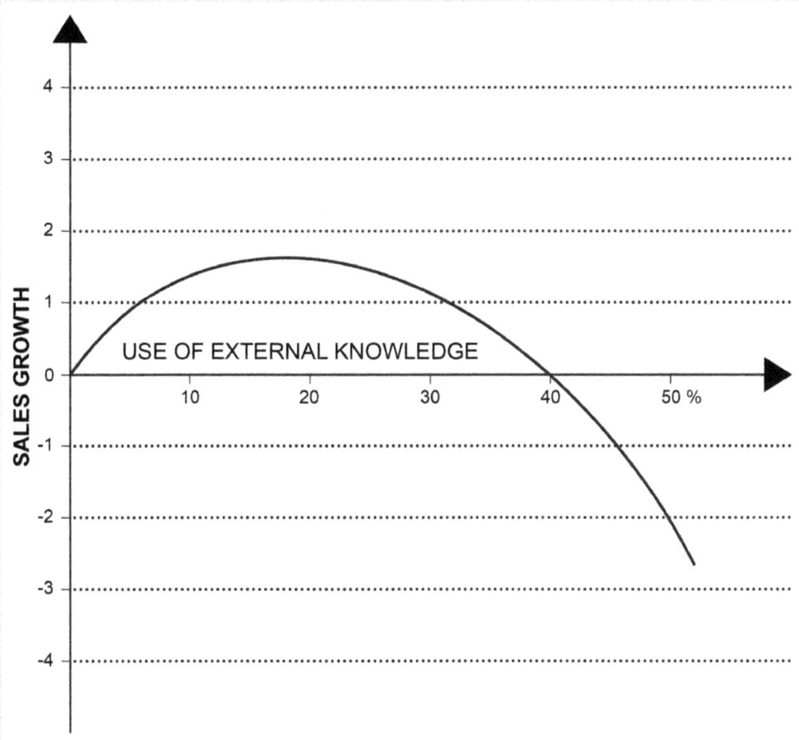

Figure 59: Formality and duration of collaboration. (Adapted with permission from Denicolai *et al.* (2016). Copyright © 2015 Elsevier Inc.)

internal knowledge and contributes to sales growth up to the relatively high level of external knowledge use of 40%. The dimension "Use of external knowledge" captures the value of inward knowledge flows, such as patents or licensing (as reported in the income statement), divided by the turnover, to provide a % figure. Beyond this point, external knowledge use has a negative effect. Internal knowledge building is rather slow but efficient in the long term. In other words, external knowledge building allows a company to absorb new knowledge quickly. Nevertheless, this quick absorption is only possible if the company has a good and well-nurtured knowledge base in a relevant area. If the internal knowledge base is too small compared to the external one, absorption fails.

5.5.2 *Joint ventures and strategic alliances*

Joint ventures usually take the form of a formal agreement between two or more organizations to codevelop a new technology or product. There are two basic types of formal joint ventures: The first is a new company formed by two or more separate organizations, which generally allocates ownership based on the share of the controlled stock. The second type is a simpler form based on a contract. The critical distinction between the two types of joint venture is that an equity arrangement requires a separate legal entity. In such cases, management is delegated to the joint venture, which does not happen with other forms of collaboration.

Unlike more formal joint ventures, a strategic alliance usually has a specific end goal and timetable, and does not normally take the form of a separate company. There are a range of motives for strategic alliances and strategies suggested to exploit each:

- to build a critical mass through cooption
- to reach new markets by leveraging cospecialized resources
- to gain new competencies through organizational learning.

In a co-option alliance, critical mass is achieved through temporary alliances with competitors, customers or companies with complementary technology, products, or services. Through co-option, a company seeks to group together other, relatively weak, companies to challenge a dominant competitor. Co-option is common where scale, or network size, is important, such as mobile telephony and aerospace. For example, Airbus was originally created in response to Boeing's dominance, while Symbian and Linux were created in response to Microsoft's dominance, and the Open Handset Alliance and Android in response to Apple's success. Greater international reach is a common motive for co-option alliances. Fujitsu initially used its alliance with ICL to develop a market presence in Europe, as did Honda using Rover. However, co-option alliances may be inherently unstable and transitory. Once the market position has been achieved, one partner may seek to take control through acquisition.

In a co-option alliance, partners are normally drawn from the same industry, whereas in cospecialization, partners are usually from different sectors. In a cospecialized alliance, partners bring unique competencies together in order to create an opportunity to enter new markets, develop new products, or build new businesses. Such cospecialization is common in systems or complex products and services. However, there is a risk associated with cospecialization: The partners are required to commit to each other's technology and standards. Where technologies are emerging and uncertain, and standards are yet to be established, there is a high risk that a partner's technology may become redundant. This has a number of implications for cospecialization alliances. First, where the dominant technologies are still uncertain, flexible forms of collaboration, such as alliances, are preferable at the early stages of an emerging market, while, at the later stages, when the market needs are clearer and the relevant technological configuration is better defined, more formal joint ventures become appropriate. Second, the implication is that the use of alliances should be restricted to instances where the technology is tacit, expensive, and time consuming to develop. If the technology is not tacit, a license is likely to be cheaper and less risky, and if the technology is not expensive or time consuming to develop, in-house development is preferable.

There are a number of reasons for the general increase in alliances and more specifically, for the switch from formal joint ventures to more transitory alliances:

- *Speed*: Transitory alliances versus careful planning. Under turbulent environmental conditions, the speed of response, learning, and lead time is more critical than careful planning, selection, and development of partnerships.
- *Partner fit*: Network versus dyadic fit. Owing to the need for speed, partners are often selected from existing network members or alternatively, due to their reputation in the broader market.
- *Partner type*: Complementarity versus familiarity. Transitory alliances increasingly occur across traditional sectors, markets, and technologies, rather than from within. An example of a complementarity partnership is the Smart Card Alliance for Near-Field Communications (NFC) mobile contactless payments, which involves banks, phone manufacturers and merchants, whereas an example of a familiarity alliance is the collaboration between BMW, Chrysler and Fiat to trial autonomous vehicles.

- *Commitment*: Aligned objectives versus trust. The transitory nature of relationships complicates the development of commitment and trust, while alliances rely more on aligned objectives and mutual goals.
- *Focus*: A few, specific tasks versus multiple roles. To reduce the complexity of managing relationships, the scope of the interaction is more narrowly defined and focused more on the task than on the relationship.

Customer integration in action: Airbus Industrie

Airbus demonstrates the benefits and challenges of joint ventures. Airbus Industrie was formed in France in 1969 as a joint venture between the German firm MBB (now DASA) and French firm Aérospatiale, to be joined by Spain's CASA in 1970 and British Aerospace (now BAE Systems) in 1979. The primary motive of this joint venture was to share the high cost and commercial risk of development. The French and German governments therefore underwrote their participation. This did not escape the attention of Boeing and the US government, which provides subsidies indirectly via defense contracts. On the other hand, all the partners had, to some extent, captive markets in the form of national airlines, although almost three-quarters of all Airbus sales were ultimately outside the EU member countries. There were also technology motives for the joint venture. For example, BAE specialized in development of the wings, Aérospatiale in the avionics, DASA in the fuselages, and CASA in the tails. However, there are now strong financial, manufacturing, and marketing reasons for combining the operations within a single company.

Airbus is not a company but a Groupment d'Intérêt Economique (GIE), which is a French legal entity that is not required to publish its accounts. Instead, the member companies absorb all the costs and any profits or losses. The partners make components according to their share of Airbus Industrie.

Before 1969, the US firm Boeing dominated the international market for civil aircraft. In the mid-80s, it accounted for 40% of the airframe market in the noncommunist world. However, the growing cost and commercial risk of airframe development resulted in consolidation of the industry and a

(*Continued*)

(*Continued*)

number of joint ventures. In addition, product life cycles had shortened due to rapid improvements in engine technology.

The Airbus partners identified an unfilled market niche for a high-capacity, short- to medium-range passenger aircraft, because more than 70% of air traffic was at that time on routes of less than 4,600 km. Thus, the Airbus A300 was thus conceived in 1969 and was essentially the result of the French and German partners, the former insisting on final assembly in France and the latter gaining access to French technology. The first A300 flew in 1974, followed by a series of successful derivatives, such as the A310 and the A320. The British partner played a leading role in the subsequent projects, bringing both capital and technological expertise to the venture. Airbus has since proved to be highly innovative with the introduction of fly-by-wire technology and common platforms and control systems in all its aircraft to reduce the cost of crew training and aircraft maintenance.

However, alliances don't always work. Table 31 shows that there are many reasons why what looks good on paper fails to work in practice!

5.5.3 *Innovation networks*

In practice, firms do not rely on a single strategy when working with partners and exploit their external knowledge. Innovation is not a solo act but a multiplayer game. Increasingly it's also about links between organizations, developing and making use of increasingly wide networks. Smart firms have always recognized the importance of linkages and connections: getting close to customers to understand their needs; working with suppliers to deliver innovative solutions; linking up with collaborators, research centers, and even competitors to build and operate innovation systems. But in an era of global operations and high-speed technological infrastructures populated by people with highly mobile skills, building and managing networks and connections has become the key requirement for innovation. It's not about knowledge creation as much as knowledge flows. Even major research and development players like Siemens and GlaxoSmithKline are realizing that they can't cover all the knowledge bases they need and therefore endeavor to build extensive links and relationships with players around the globe.

Table 31: Common reasons for alliance failures (review of 16 studies) derived from Duysters *et al.* (1999).

Reason for failure	Studies reporting factor (%) (n = 16)
Strategic/goal divergence	50
Partner problems	38
Strong–weak relation	38
Cultural mismatch	25
Insufficient trust	25
Operational/geographical overlap	25
Personnel clashes	25
Lack of commitment	25
Unrealistic expectations/time	25
Asymmetric incentives	13

Networking is important right across the innovation process — from finding opportunities to pulling together the resources to develop the venture, to making it happen and diffusing the idea — to capture value at the end of the process. The idea is not new. Networking of this kind is something which Roy Rothwell, for many years a key researcher at Sussex University's Science Policy Research Unit (SPRU), foresaw in his pioneering work on innovation models. This study predicted a gradual move away from thinking about (and organizing) a linear science or technology-push or demand-pull process to one which saw increasing interactivity (Rothwell, 1992). At first, this process exists across a company with cross-functional teams and other boundary-spanning activities. Further generations represent higher extents of external participation. Rothwell's vision of the "fifth generation" innovation is essentially the one in which we now need to operate. It has rich and diverse network linkages accelerated and enabled by a concentrated set of information and communication technologies (Table 32).

Procter and Gamble is a good example of a company that made this shift in approach. In the late 1990s, there were concerns about its

Table 32: Rothwell's five generations of innovation models (Rothwell, 1992).

Generation	Key features
First/second	Simple linear models — need pull, technology push
Third	Coupling model, recognizing interaction between different elements, and feedback loops between them
Fourth	Parallel model, integration within the company, upstream with key suppliers and downstream with demanding and active customers, emphasis on linkages and alliances
Fifth	Systems integration and extensive networking, flexible and customized response, continuous innovation

traditional inward-focused approach to innovation. Although this approach worked, there were worries related to the rapidly rising costs of carrying out R&D. Additionally, there had been many innovations which the company might have undertaken but did not, only to find someone else doing so and succeeding. The CEO Alan Lafley explained, "Our R&D productivity had levelled off, and our innovation success rate — the percentage of new products that met financial objectives — had stagnated at about 35%. Squeezed by nimble competitors, flattening sales, lackluster new launches, and a quarterly earnings miss, we lost more than half our market."

Customer integration in action: Connect and develop at Procter and Gamble

P&G's successes with "connect and develop" owe much to it mobilizing rich linkages between people who know things within its giant global operations. Amongst its internal networking successes was the Crest Whitestrips product, which essentially linked oral care experts with researchers working on film technology and others in the bleach and household cleaning groups. Another was the Olay Daily Facials, which linked its surface-active agents' skin care expertise with people from the tissue and towel areas and from the fabric property-enhancing skills developed in "Bounce," a fabric softening product.

(*Continued*)

(*Continued*)

Making this type of innovation happen as part of daily life rather than as a special initiative is a big challenge. P&G uses multiple methods, including extensive networking via an intranet site called "Ask me," which links 10,000 technical people across the globe. This intranet site acts as a signposting and web market for ideas and problems across the company. P&G also operates 21 "communities of practice" comprising people with key areas of expertise, such as polymer chemists, biological scientists, people involved with fragrances. The company also has a global technology council, which is made up of representatives of all of its business units.

All of the above comprised the first step toward opening up the company's R&D. The next step included opening beyond the company premises. By 2006, 35% of new products comprised ideas, or other elements, originating outside P&G compared with only 15% in 2000. In the past 2 y, more than 100 new products have been introduced by mobilizing external innovation networks. While external collaborators are often researchers and technical experts, collaboration also includes ordinary consumers. It differs from simple outsourcing, which means hiring people to perform a specific task. The OI approach is more about entering a relationship and involving outsiders in creative, collaborative processes. P&G's ability to find relevant partners, form a network with them, and ultimately operate that network is a key competence for this exchange to work. Professional IP management, building trust, sharing key information, and avoiding "free riders" are part of this success.

We can map some of these different types of innovation networks in a simple diagram (Figure 60) which positions them in terms of:

- how radical the innovation target is with respect to current innovative activity
- the similarity of the participating companies.

By making a distinction between the degree of newness to which the innovation agenda aspires and the innovation networks' degree of heterogeneity, Figure 60 visualizes different forms of innovation networks' value for resolving different issues. In zone 1, we have individuals and

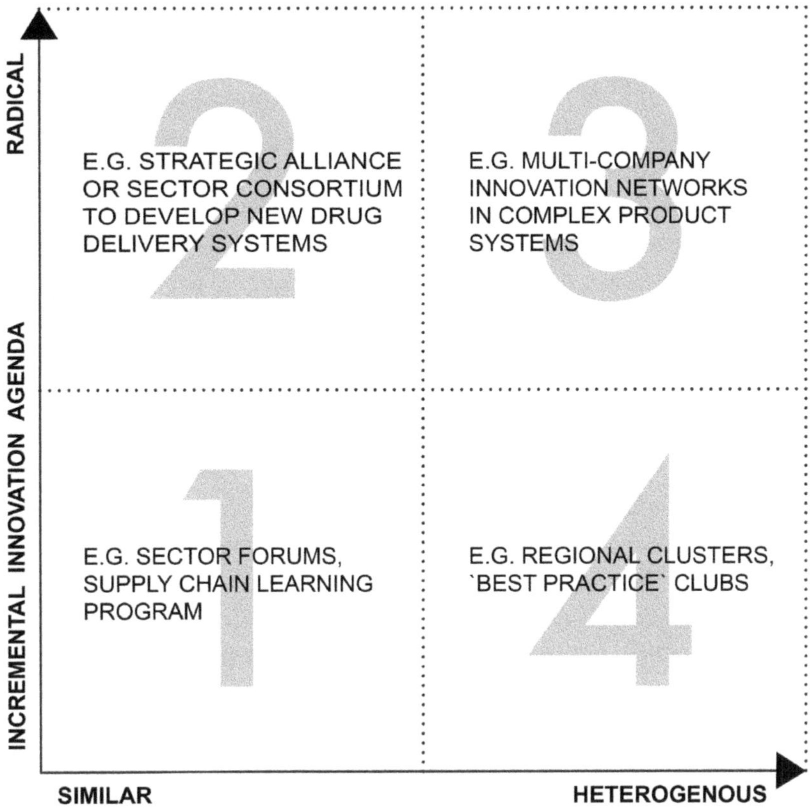

Figure 60: Types of innovation networks. (Reproduced with permission from Tidd and Bessant (in press).)

organizations with a broadly similar orientation working on tactical innovation issues. This might be a cluster, or sector forum, concerned with adopting and configuring "good practice" manufacturing, or a group of innovation managers in the health sector trying to improve productivity. Issues here would involve enabling them to share experiences, disclose information, develop trust and transparency, and build a system-level sense of shared purpose based on the innovation. For example, the Aerospace Supply Chain Excellence Programme includes 40 suppliers to Airbus, including BAE Systems and Rolls-Royce, and aims to promote closer coupling between customers and suppliers and collective work by suppliers such as Shared Services and Extended Enterprise.

Zone 2 activities might involve players from a sector working to explore and create new product, or process, concepts, for example, the emerging biotechnology or pharmaceutical networking regarding frontier developments and the need to look for interesting connections and synthesis between these adjacent sectors. Here, the issue is exploratory, thus challenging the existing boundaries, but it will rely on a degree of information sharing and shared risk-taking, often in the form of formal joint ventures and strategic alliances. In financial services, banks are investing in and working with suppliers of technology, so-called fintech firms, to develop new services, for example, BBVA bank's relationships with fintech suppliers Simple and Holvi to develop online banking.

In Zone 3, the players are highly differentiated and bring different key pieces of knowledge to the party. The risk that they may disclose sensitive information is high, so ensuring careful IP management and establishing the ground rules are crucial. Since this kind of innovation is likely to involve many other risks, putting risk- and benefit-sharing arrangements in place is also critical. For example, Honda Motor Co Ltd and Hitachi Ltd formed a joint venture in 2017 to develop, produce and sell motors for electric vehicles, crossing the automotive and electronics industries.

Zone 4 involves the kind of shared learning across organizations which we saw earlier — essentially building on regional or sectoral links to focus a shared learning effort. Zara is a good example, as its fast-fashion strategy is based upon rapid responses from its European regional manufacturing networks, rather than more conventional retail supply networks in Asia, which are cheaper but slower.

Research note: High-value innovation networks

In a review of innovation networks in the UK, researchers from the Advanced Institute of Management Research (AIM) found that the following characteristics were important success factors:

- *High diversity*: Network partners from a wide range of disciplines and backgrounds who encourage exchanges about ideas across systems

(*Continued*)

(*Continued*)

- *Third-party gatekeepers*: Not only science partners, such as universities, but also consultants and trade associations, who provide access to expertise and act as neutral knowledge brokers across the network
- *Financial leverage*: Access to investors via business angels, venture capitalists, and corporate venturing, which spreads the risk of innovation and provides market intelligence
- *Proactive management*: The participants regard the network as a valuable asset and actively manage it to reap the innovation benefits.

Source: Advanced Institute of Management Research (2004).

Building networks involves three steps: finding, forming, and performing. The "finding" stage is essentially about setting up the network. The key issues here are providing momentum to bring the network together and defining its purpose clearly. This stage may be crisis triggered, for example, a perception of an urgent need to catch up via innovation adoption. Equally, a shared perception of opportunity — the potential to enter new markets or exploit new technologies — may drive this stage. Third parties, such as network brokers, gatekeepers, policy agents, and facilitators, will often play key roles during this stage.

"Forming" a network involves building a type of organization with some structure to enable its operation. The key issue here is trying to establish core operating processes (which are supported and agreed upon) to deal with the following:

- network boundary management — how the membership of the network is defined and maintained
- decision-making — how (where, when, who) decisions are made at the network level
- conflict resolution — how conflicts are effectively resolved
- information processing — how information flows between members and is managed
- knowledge management — how knowledge is created, captured, shared, and used across the network

- motivation — how members are motivated to join, or remain within the network
- risk or benefit sharing — how the risks and rewards are allocated across the network members
- coordination — how the network operations are integrated and coordinated.

There is no universal best way of governing networks. For example, the methods and the intensity of the communication and knowledge sharing will differ in a project's exploration and exploitation phases (Sørensen and Mattsson, 2016).

Finally, the "performing" stage is about operating the network and allowing it to evolve. Networks need not last forever; sometimes they are set up to achieve a highly specific purpose (e.g., the development of a new product concept), and once this has been done, the network can be disbanded. In other cases, the networking activities can be sustained for as long as the members see benefits. This may require a periodic review and "retargeting" to maintain motivation. Shared values and expectations promote commitment to a network but are not sufficient to produce innovation outcomes, which is more a function of network size and density (Hammarfjord and Roxenhall, 2017).

Innovation networks are more than just ways of assembling and deploying knowledge in a complex world. They can also have "emergent properties"; that is, the potential for the whole to be greater than the sum of its parts. Being part of an effective innovation network can deliver a wide range of benefits beyond the collective knowledge efficiency mentioned above. These benefits include gaining access to different and complementary knowledge sets, reducing risks by sharing them, accessing new markets and technologies, and pooling complementary skills and assets. Without such networks, it would be nearly impossible for a single company, or lone inventor, to bring his or her idea to market successfully. The need to extend their access to key innovation knowledge and resources is the main reason why established businesses are increasingly turning to cooperation and alliances.

Choosing the most appropriate form of innovation network is very much dependent on where an organization's size, sector, and life cycle stage are. For example, customer inputs are central when market dynamics are high, suppliers are important in technologically challenging environments, and including companies from other industries is necessary for more radical innovation. Table 33 identifies the potential benefits and challenges of network approaches.

Table 33: Potential benefits and challenges of applying OI. (Reproduced with permission from Tidd (2014). Copyright © 2014 Imperial College Press.)

Six principles of OI	Potential benefits	Network challenge
Tap into external knowledge	Increase the pool of knowledge Reduce reliance on limited internal knowledge	How to search for and identify relevant knowledge sources How to share or transfer such knowledge, especially tacit and systemic knowledge
External R&D has significant value	Can reduce the cost and uncertainty associated with internal R&D and increase the depth and breadth of R&D	Less likely to lead to distinctive capabilities and more difficult to differentiate External R&D also available to competitors
Do not have to originate research in order to profit from it	Reduce costs of internal R&D, more resources for external search strategies and relationships	Need sufficient R&D capability in order to identify, evaluate, and adapt external R&D
Building a better business model is superior to being first to market	Greater emphasis on capturing rather than creating value	First-mover advantages depend on technology and market context Developing a business model demands time-consuming negotiation with other actors
Best *use* of internal and external ideas, not the *generation* of ideas	Better balance of resources to search and identify ideas, rather than generate	Generating ideas is only a small part of the innovation process Most ideas are unproven or have no value, so the evaluation and development costs are high

(*Continued*)

Table 33: (*Continued*)

Six principles of OI	Potential benefits	Network challenge
Profit from others' intellectual property (inbound OI), and others' use of our intellectual property (outbound OI)	IP's value is very sensitive to complementary capabilities such as brand, sales network, production, logistics, and complementary products and services	Conflicts about commercial interests or strategic direction Negotiation of acceptable forms and terms of IP licenses

Chapter 6

Development of organizational customer integration capabilities and implementation of customer integration in NPD and NSD

6.1 Three process-related customer integration competencies and the customer orientation mindset

An innovating company needs to have customer integration competency to fully profit from customer integration activities. Competencies are "a set of integrated, harmonized abilities that distinguish a firm in the market place" (Holahan *et al.*, 2014, p. 331). In combination with other corporate competencies, customer integration competencies provide the basis for competitive advantage and innovation success to offer superior customer value, differentiate a company's offer from that of competitors, make it difficult to imitate, and open up the organization to new markets.

Customer integration competency includes a bundle of capabilities that allow a company to learn from interactions with customers and to create an appropriate value based on this learning. These capabilities guide the acquisition, dissemination, and utilization of customer information (Fiol and Lyles, 1985; March, 1991; Kok *et al.*, 2003) (Figure 61).

Customer information acquisition capabilities require knowledge of customer involvement tools, ways of selecting and attracting appropriate

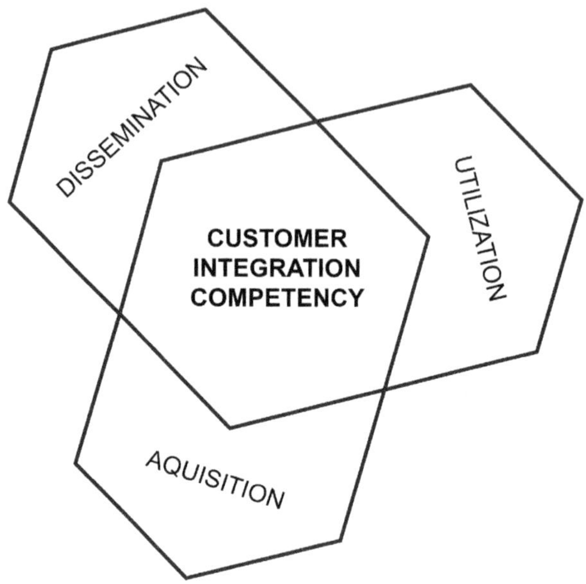

Figure 61: The interplay of capabilities that create customer integration competency.

customers, and practical experience with applying the tools. Customer information dissemination capabilities refer to all the capabilities needed to internally and externally distribute the insights gathered through customer information acquisition practices. In this context, knowledge is relevant regarding whom to include when during the customer involvement process and how to transform and edit the information to allow the information recipient to digest it easily. The internal entities that receive the distributed knowledge need customer information utilization capabilities to apply them to fulfill their tasks. These capabilities require willingness and ability to absorb customer information and to interpret and use it meaningfully.

In this context, we found that many companies start customer integration initiatives with varying degrees of success but only a few take the important step toward institutionalizing their customer integration efforts in order to ingrain them as a permanent element of the NPD process. This institutionalization process is a long-term internal learning process that requires nurturing. However, only this institutionalization allows a company to fully profit from its customer integration by building the absorptive capacity required to appreciate and utilize customer input appropriately. Gathering external input from customers through direct or indirect customer integration methods can rapidly develop knowledge for a specific project. However, this knowledge relies on exploiting existing capabilities in order to utilize this external customer input for rather immediate benefits. Extending existing capabilities through an institutionalized customer integration approach allows for building more sustainable, self-renewing, and longer-lasting benefits. Such an endeavor requires investments in people and firm practices that allow for growing an integrated, functioning system that combines internal NPD processes with customer integration methods. The strength of an institutionalized approach lies in its ability to accumulate and consolidate customer integration method knowledge and customer integration results over time and to manage knowledge flow from one project to future projects. The initial ability to learn from customers is constrained in noninstitutionalized customer integration initiatives, because they act in isolation. An institutionalized approach helps companies build a complete picture of their customers and profit from multiple integration approaches and multiple sources instead of building on isolated initiatives. In a nutshell, a noninstitutionalized approach to customer integration allows for utilizing absorptive capacity, while an institutionalized process allows for building absorptive capacity.

Research note: The magic and burden of absorptive capacity

Absorptive capacity is the ability to obtain, assimilate, and exploit information (Cohen and Levinthal, 1990). In order for an absorbing unit to absorb new information on a specific area, for example, the information from external partners such as customers, it has to possess a fraction of knowledge about this area. This existing fraction not only helps it comprehend chunks of new information but also influences the way in which the information is understood and interpreted. Underdeveloped knowledge hinders absorption. A high degree of knowledge is also not beneficial, as it reduces the chances of finding new knowledge through an external search.

(*Continued*)

(*Continued*)

The capability to absorb and exploit external inbound knowledge, as well as to understand how internal R&D can facilitate or hinder this, is an important challenge. A study of 325 firms over 5 y found that external knowledge is more expensive to acquire than internal knowledge, but that its deployment can be much faster if previous internal R&D investment supports this (Denicolai *et al.*, 2016).

In other words, there is an inverted curvilinear relationship between existing knowledge and new knowledge's absorbability. This relationship hampers the successful absorption of information for products that are radically new for the innovating company. Per definition, an innovating company's absorption capacity is really low in such radical NPD situations. For example, this might be the case if a railway company were to explore ideas related to unmanned aerial vehicles (UAVs) for the railway system even though they are a distant expertise area outside the company's core competencies. In such a case, the innovating company will incur problems with recognizing relevant developments in the markets and technologies and with judging the potential value of idea propositions. Facing the task of evaluating ideas, companies tend to ignore ideas from highly experienced individuals, as they provide very specific solutions that the company cannot understand. Instead, they rather rely on quite general propositions from less experienced individuals, because companies lacking absorptive capacity find these ideas are easier to understand. Promising ideas are therefore often ignored because they are not understood. A potential way out of such a dissatisfactory situation involves using innovation intermediaries to evaluate the relevant solutions. For example, innovation intermediaries with algorithm-based big data search techniques that can provide an innovating company with need-related knowledge of external contributors can accelerate access to knowledge and speed up the absorption capacity building (Kokshagina *et al.*, 2017).

In this context, calling on customer information is a promising approach for simultaneous consideration of need-solution pairs that enables a more appropriate evaluation of radically new concepts and reduces the risk of eliminating potentially interesting solutions due to lack of absorptive capacity (von Hippel and von Krogh, 2017).

6.2 Customer information acquisition capabilities

An innovating company needs to develop a sound knowledge and practice of applying diverse customer integration tools for specific goals. Building knowledge on existing methods and keeping up to date with new ones are central. Further, an innovating company needs to understand the way in which these methods can be applied in a specific context and which method is the most appropriate for which innovating problem.

Besides method competence, customer information acquisition capabilities also require savviness with selecting and attracting contributors. Contributors include customers, employees, and third parties. In respect of customers, it is important to think about the type and quantity of customers required for the successful execution of a specific customer integration method. For example, several methods for opening up the innovation process, such as the lead user method, call for customers with specific domain-relevant skills. Many methods to close the solution space require a representative sample of ordinary users that is large enough to draw conclusions regarding the general target population. Once the right type of customers has been identified, the innovating company has to consider the best way to reach them and to motivate them to participate. During and after the customer integration project, the following are essential: understanding motivation, deciding on the benefits and the incentives required to stimulate participation, understanding the participants' satisfaction with the customer integration experience, and the ways to reduce customers' opportunistic retaliation and other counterproductive behavior. Monetary rewards are not always the best choice (Benkler, 2002; Lettl, 2007).

In respect of employees, it is important at an early stage of the customer integration project to clarify which employees should later work with the results, because integrating them early will allow them to influence the data acquisition, thus ensuring the data's meaningfulness for them, while being actively involved will encourage them to actually use the generated data later. In this respect, it is important to understand what a particular individual or department might win or lose if a certain innovation project is realized. Fear of losing power, or relevance, may make some entities reluctant to embrace customers' innovation ideas, even though they are promising for the company in general. Given this fear of losing relevance it is, for example, no wonder that, in 2012, Beth Wozrick, president of the environmental and combustion controls at Honeywell, mocked the idea of self-learning controls and held that consumers prefer to control thermostats rather than thermostats controlling them (Verganti, 2016). Idea evaluations can only be critical and valuable if the evaluator is not biased against ideas for personal career reasons or similar issues. In other words, companies need the capability to be critical. Top management has to develop a clear understanding of which customer insights are relevant but cannot be realized within a certain business unit due to power plays, fears about losing influence, and similar reasons. In respect of such customer insights, the top management should consider alternative, neutral units or spin-offs that facilitate realization.

Besides integrating employees and customers, it is worth thinking about integrating third parties. Third parties can be intermediaries that help a company connect with the right customers (e.g., Innocentive, or HeroX, which can provide access to a large pool of potential problem solvers in order to run successful idea competitions), firms that can carry out a customer integration approach professionally (e.g., consultancies like Hyve, or Strategyn, which specialize in specific customer integration methods), firms that can be ecosystem partners (e.g., suppliers, trade agents), and other types of stakeholders (e.g., pressure groups, legal institutions) whose cooperation is required to launch an innovation.

Furthermore, successful information acquisition through customer integration projects requires the project goals and expected output to be defined accurately and concisely. In this context, an innovating company has to set a clear goal, ensure that (1) it reveals the right amount of relevant information about the project, (2) this information is communicated well enough for the customer to understand the purpose of their participation, and (3) they can contribute meaningfully to it. Closely related to information disclosure is a clear strategy to protect the result of the customer integration process.

Information leakage to competitors by customer's or by competitors' access to ICT-enabled customer integration tools is a legitimate fear that could deter an innovating company from making disclosures. Other potential threats related to opportunistic customer behavior take the form of B2B customers turning into future competitors or idea-contributing customers taking legal action to claim higher than the agreed upon contribution compensation. Generally, rules and contractual provisions in the form of nondisclosure agreements bar customers from sharing information on the customer integration project with any third entity. Clear agreements on entitlement and royalties for idea contributors are also essential (Piller and Ihl, 2009; Pera *et al.*, 2016).

In order to run customer integration projects effectively and efficiently, it is important to have a person, or unit, within the innovating company with a clear responsibility for coordinating such projects. New technologies, constantly changing environmental contexts, and evolving customer integration methods continuously open up new opportunities and challenge the existing ways of integrating the customer. The entity responsible for acquiring customer information must therefore have a sound understanding of and experience with existing customer integration methods, as well as the willingness and ability to constantly broaden its knowledge and to apply and adapt existing knowledge to unfamiliar and new circumstances.

6.3 Customer information dissemination capabilities

Customer information dissemination is relevant for an innovating company when the entity gathering customer information and the entity using it are not the same. Large companies maintain customer insights departments whose role is to gather customer information. These departments follow top-management-driven customer insights projects and projects that other departments initiate, such as strategic business units, which are often the ones responsible for innovation processes in B2C companies, and R&D departments, which are often the ones responsible for innovation processes in B2B companies. When the customer insights department has completed its work, it presents the results and hands over a report on the findings to the business unit or R&D department. While strong visualization skills are an advantage to summarize results effectively, they can never live up to actual experience. Through its direct interaction with customers, the customer insights department builds a fine-grained understanding of the customer; consequently, reducing these insights to a report inevitably leads to a knowledge gap between the customer insights department and the department using the information to develop an innovation further.

In our work with companies, we found that this knowledge gap, enriched with a decision gap, is a key sign of a lack of customer information dissemination capabilities. In this context, a decision gap means that the customer insights team's responsibility for the customer integration project ends once they have revealed the results. Further activities and decisions based on these insights are left to the unit that initiated the project in the first place. A customer insights manager at a leading electronic consumer company mentioned: "Sometimes I think we need to stay in the process longer. Since we are a big company, everyone has a particular field of expertise — everyone just does their little part. A business approaches us, we do research, we conclude this with a report, and then we are excluded. I do not have any power to influence the business. It would be good to be there to raise my voice, but sometimes they do not even invite me to the meetings where the decision is made, because I might start asking difficult questions."

Add a timing gap to the knowledge and decision gap, and you have a full picture of the challenges of information dissemination: In some companies, a customer integration project can only be initiated for innovation projects for which the innovation managers can demonstrate a profitable business case proving that the innovation is worth an investment. Building a business case necessitates a set of assumptions about the customer and the product that already provides a quite concrete idea for solving a specific customer need. An innovation manager recently told us that his boss's mantra is: "I am not going to let you investigate until you have something to investigate," which implies that the innovation manager already needs to have a specific product or service in mind before talking to the customer.

Such a procedure reduces the value of certain methods, notably of indirect approaches to open the solution space. Late customer integration to detect latent needs signifies that the innovation team had a product or idea in mind first and then endeavored to find a need for this. Instead of trying to understand the users better, the team looks for ways to validate an existing solution. Proactive customer orientation should work the other way round: detecting a latent need and then generating ideas for solutions that satisfy this need.

The groundwork required for a business case also shapes the business unit team's viewpoint. When the time comes to finally commission a customer integration project, the team already has a preconceived view of what the customer should think, say, and like. Consequently, the groundwork may reduce the business unit team's readiness to abandon an idea if customers do not assess it positively for both psychological and structural reasons, because at that stage, the team has already "promised" the management team to deliver the product and has included it in the roadmap. The situation is similar for an R&D team: If it has already invested time and money in the technical realization, is emotionally attached to a technical detail that it helped solve, or simply does not want to miss deadlines, such an R&D team might be reluctant to make significant changes to technical specifications, even if customer insights seem to require these. Berchicci and Tucci (2010) showed that this may even lead an innovation team to ignore certain customer arguments, because these customers do not share their values and criticize the product. Ultimately, it may mean overlooking valuable improvement suggestions.

Small collaborating teams in which the team members who gather information work side by side with those members utilizing the information reduce the dissemination problem within this unit. Nevertheless, customer information dissemination across units is more difficult than in organizations where one unit is responsible for customer information gathering on all projects, and another unit uses this information. In other words, small cross-functional units hamper the beneficial effects of synergies and reduce double work across organizations. A matrix organization that disseminates customer information is appropriate to reduce such countertendencies and to use the benefits of customer information dissemination within and across units. In such an organization, the employees do information gathering work in interdisciplinary innovation teams some days and, in other days, work in the specialized customer information gathering departments.

Customer integration in action: Unilever's customer intelligence approach

At Unilever, the market research department developed from a reactive unit supplying reports that the marketing department requested into a strategically involved partner focusing on interpreting data. The unit developed an IT-based global marketing information system that aggregates large chunks of customer-related data from multiple internal (e.g., product sales figures, spending on media, and call-center records) and external sources (e.g., social media monitoring) and ensures all other departments have access to this system. While the finance and the marketing departments used to have different figures before, the new system allows a consistent view of the data and reduces debates on the definition of the data and the methodology. It also reduced the market research department's work on reports, freeing time to distill the insights and develop recommendations.

A cholesterol-lowering drink was one of the insights that the market research department generated. This drink has to be used consistently to have an impact. The department found that consumers needed a minimum of 3 weeks to ensure that their consumption pattern would change to consistent usage. It also revealed that such a change works best if peer pressure is created in group work. This led to the development of the "It takes a village" campaign. The program runs successfully in communities and town in 10 countries and motivates each town to fully engage in lowering their inhabitants' cholesterol. Unilever provides cholesterol-testing opportunities, as well as nutritional and exercise advice.

The campaign has helped more than 80% of the participants lower their cholesterol and has increased Unilever's product sales.

Source: van den Driest *et al.* (2016).

Established processes and routines for initiating customer integration projects and disseminating the acquired customer data strongly impact the value derived from customer integration projects. Being aware of the way in which information is passed from one party to another and adapted during the process is a first step toward understanding the internal dissemination processes of customer information. The next step involves challenging existing practices and improving them constantly, because they influence the way and extent to which a customer integration project's results are finally integrated into and used in the NPD or NSD process (Gebauer *et al.*, 2012; Holahan *et al.*, 2014).

When it comes to the knowledge dissemination of customer insights, companies are usually so busy ensuring that the internal processes are correct that they forget to consider ways of sharing their customer insights with their business partners. A clear external dissemination strategy should clarify which information is shared with which supplier, R&D partner, trade agent, or other business partner. Such a strategy includes clearly addressing the goal of sharing this information. It should also address types of information that are not shared and the reasons for not sharing them. Such a strategy helps employees exchange appropriate information in their daily interaction with their business partners.

Whatever the case, successful dissemination needs clear and well-thought-through processes, an associated IT infrastructure, and an entity responsible for coordinating the dissemination. It is not ideal for this entity to be the same as the one responsible for gathering the customer data. The latter function requires in-depth expertise on tools and methods, while a generalist entity with strong interpersonal skills to run the flexible, but disciplined, processes of transferring information successfully should hold the former function. These interpersonal skills include negotiation, moderation, and conflict management skills. This entity has to encourage others to share and discuss the options, their views, and interpretations of the customer information. Creative criticism inspires the subsequent processing and use of the insights provided by customer integration projects, as it helps innovation teams question their assumptions, reflect on the innovation, and make appropriate changes. The right balance between tensions and harmony has to be found if customer integration projects are not to end in disastrous and insurmountable turf wars.

6.4 Customer information utilization capabilities

Analytical interpreting and selection skills help utilize customer information to the innovating company's advantage. Once it starts, the innovating company will gather a large amount of information through customer integration. Direct and indirect customer integration approaches to open the solution space provide an influx of new product or service ideas and concepts. Evaluative feedback on new product or service ideas and concepts will come from customer integration approaches that close the solution space. Some of this information will seem inconsistent, vague, or ambiguous. NPD/NSD teams will never be able to gather all the information required to take decisions with full certainty. NPD/NSD teams need to easily deal with paradox and ambiguous situations. They should be able to turn uncertain situations in their favor by looking for new opportunities and any valuable information. At the same time, they need to be ready to look beyond such situations' surface. Finding that one customer wants this and another the opposite should not lead to the conclusion that customers themselves do not know what they want and that gathering customer insights is meaningless. Instead, such results should stimulate further thoughts about the reasons for differences. For example, customers' differing needs may simply be due to them belonging to different market segments that require different offerings. The NPD/NSD team should strive to analyze the different segments' characteristics and attractiveness to determine which the customers want to satisfy.

Research note: The effect of decision-maker orientation and diversity on the ability to utilize customer information

The NPD/NSD team's cognitive styles and diversity influence their readiness to utilize customer information that challenges current practices and perspectives in terms of opening up to radical innovation.

Regulatory focus theory (Higgins, 1997, 1998) describes two processes through which individuals align themselves with external standards and targets. One is a promotion-focused process that refers to individuals striving for achievements and motivated by of rewards. This focus encourages individuals to find ways to realize their hopes and aspirations, be creative problem solvers, take risks, and be on the lookout for change and novelty. The other is a prevention-focused process enticing individuals to zoom in on the avoidance of punishment and to value safety and security highly (Crowe and Higgins, 1997; Kark and Dijk, 2007). Innovation team members with a prevention attitude are more risk averse and likely to be skeptical about the radical ideas that consumers may propose.

(*Continued*)

(*Continued*)

Customers can sometimes propose totally radical ideas that are difficult to implement due to the companies lack of relevant competencies and capabilities. Having such team members evaluate radical customer ideas is likely to result in idea rejection. Promotion-oriented team members are in a better position to embrace new, radical ideas that challenge status quo knowledge. Overrelying on radical ideas might, however, lead to products that are too removed from the current market realities and might only cater to a small innovative target group's the urgent needs but are irrelevant for the majority of customers. A mix of team members with preventive and promotion-oriented attitudes will constructively, and from different angles, reflect on the risks of and chances offered by radical customer ideas. Thus, the team should provide the diversity required to develop such ideas further until they are fully fledged, persuasive, and unique market offerings.

In a similar vein, studies on team members' cognitive styles have found that creative and conformist members are more open to radical ideas, while the opposite holds true for attentive-to-detail members. Creative members enhance task conflicts and hinder team adherence to standards. However, teams dominated by creative members are not more innovative than average. Instead, the ideal team composition for radical innovation is around 20–25% creatives, plus conformists, and attention-to-detail members (Miron-Spektor *et al.*, 2011).

Upper echelon theory argues that the idiosyncrasies of decision-making teams influence strategic choices. Team diversity due to the heterogeneity of its members' educational background, functional background, industrial background, organization background, and board tenure encourages the specification of innovation fields. More diverse teams are more likely to have opposing views. As a consequence, they hold more discussions on the right customers and technologies to target, resulting in the establishment of long-term innovation fields and a clearer specification of the customer needs and the groups on which the company plans to focus its technologies within these fields. More diverse teams are therefore in a better position to carry out longer-term innovation projects focusing on more radical ideas (Talkea *et al.*, 2010; Østergaard *et al.*, 2011).

In direct customer integration approaches to open the solution space, the number of submitted ideas is usually quite high, making it difficult to properly analyze the full potential of each idea. Customer-developed ideas might be presented in a form that addresses unmet consumer needs, but the corporate idea evaluation committee might not find this link easy to comprehend, which means the opportunity to make use of an idea may be lost. This situation occurs if the proposed ideas are too distant from what the company is currently doing (in terms of the existing markets or technologies) and do not match the company's internal capabilities. Recent studies have found that the probability of an innovating company's idea evaluators ignoring distant knowledge increases with the number of ideas the evaluators have to evaluate. In order to save time, such evaluators resort to easily available knowledge and filter out ideas from unfamiliar knowledge areas, resulting in the selection of incrementally new ideas at the detriment of radical ones (Piezunka and Dahlander, 2015; Criscuolo *et al.*, 2017).

Further, some ideas can be formulated too vaguely and in too few details. This might result from not assessing these ideas, although it might be worth further developing them in order to transform them into more convincing propositions (Nishikawa *et al.*, 2013). Le Masson and Magnusson (2002) suggest that evaluators should look at ideas from a broader perspective, instead of simply assessing a particular idea on its own. This broader perspective involves clustering, combining, and reconfiguring simple consumer ideas into attractive constellations by creating connections and elaborating on a set of vague, or not-yet-feasible interesting ideas. In other words, single ideas may be considered irrelevant, or not powerful enough on their own, but they can turn into impactful propositions once analyzed and grouped with other ideas.

In indirect approaches to open the solution space, identified latent needs are the grains of innovation. The way in which the innovating company uses latent needs information determines whether the grains grow into a successful new offering. The tendency to look at existing clients, or potential future customers (see Section 1.6 on proactive and responsive customer orientation), specifically influences the way in which the innovating unit transforms latent needs into product offerings.

In approaches to close the solution process, the understanding of customer orientation within the innovating company also impacts the way it utilizes customers' evaluative feedback. Innovation teams have a tendency to ignore, or downplay, information originating from sources with a mindset that largely differs from their own and if this information contradicts their point of view. Current customers' positive feedback on incremental improvements is thus more likely to be utilized than potential new customers' critical feedback on more radical offerings. This tendency might result in companies incorporating a responsive customer orientation filtering out radical innovation. For companies with a proactive customer orientation, the tendency might result in the ignoring of relevant critical information on radical concepts, which could eventually result in radical innovation endeavors having little success (Berchicci and Tucci, 2010). Proactive companies need to develop market visioning competence and future market sensibility to utilize customer feedback information profitably.

Innovating companies with a future market vision build a clear and specific image of an important future market for a radical innovation (Reid and de Brentani, 2010, 2015). Utilizing potential future customer feedback combines market visioning competence with the ability to use negative customer feedback on radical ideas constructively and to identify the most attractive future opportunities and combine it with knowledge of the innovating company's strategic strengths and the gathered customer insights.

6.5 The way ahead

Markets and technologies are constantly changing, seemingly happening at an increasingly accelerated pace. These developments offer huge opportunities. Companies need to be agile and able to strengthen and reconfigure their capabilities dynamically to respond to and actively drive the opportunities that new markets and technologies offer. Simultaneously, companies need to be strong, concise, and consistent in their strategic actions. Strong leaders understand that an active exchange of information between the company and external sources allows for balancing the countertendencies between flexibility and consistency. Customers, who are understood as the current clients, and future prospects are among the most central external sources whom the company should engage with.

Some companies embrace customer integration and move faster, others follow. We observe differences in the successful use of customer integration for NPSD — not only across organizations but also within organizations. These differences are rooted in a different willingness and ability to test customer integration approaches.

Customer integration in action: On institutionalizing customer integration thinking

In a 2016, BCG's Center for Customer Insight survey revealed that 75% of top managers believe that the professional use and handling of customer insights are pivotal for business growth and prosperity. However, the majority of companies still only use selective customer integration tools on a tactical level (Barton *et al.*, 2016). The degree to which customer information is retrieved and utilized in innovation projects is often left to the project manager's discretion. While individual initiatives are an important start, finding followers, implementing a sound customer integration process throughout the whole organization, and linking the process clearly to the strategic foresight process are central for long-term success.

Understanding customer integration methods and their applications is the first step toward improving a company's chances to use these methods efficiently and effectively. This book offers the basics for this central step. The next step is for companies to test the methods and learn from this experience. When using a method for the first time, it is advisable to consult a specialist with method competence in the specific customer integration method. With increasing method competence, companies can start adapting and combining methods to fit the specific needs of certain NPD/NSD projects or processes. Further, it makes sense to start customer integration approaches in projects whose project leaders are not open to these approaches, to delivering success stories, to attracting ambassadors, and to scaling up at a later stage. Striving stubbornly to convince people who hold a traditional view is usually not a typical way to successful customer integration. A consistent customer integration process that the whole organization supports takes time, because the customer orientation and mindset have to fit and the customer integration knowledge and skills have to be built.

We conclude with a Senegalese wisdom: "The giant tree grows from a grain." Nurture the tree, but remember that haste and hurry will not make the tree grow faster or better. The tree will eventually become strong and fruitful, and birds will flock to it.

References

Abetti, P.A. (2000). Critical success factors for radical technological innovation: a five case study. *Creativity & Innovation Management*, **9**, 208–221.

Advanced Institute of Management Research. (2004). I-Works: How High Value Innovation Networks Can Boost UK Productivity. Executive briefing. Advanced Institute of Management Research, London.

Afuah, A. (2000). How much do your co-opetitors' capabilities matter in the face of technological change? *Strategic Management Journal*, **21**, 397–404.

Afuah, A., Tucci, C.L. (2012). Crowdsourcing as a solution to distant search. *Academy of Management Review*, **37**, 355–375.

Agogué, M., Levillain, K., Hooge, S. (2015). Gamification of creativity: exploring the usefulness of serious games for ideation. *Creativity & Innovation Management*, **24**, 415–429.

Akao, Y. (2004). *Quality Function Deployment: Integrating Customer Requirements into Product Design*. New York, NY: Productivity Press.

Alam, I. (2006). *Process of Customer Interaction in New Service Development in Involving Customers in New Service Development*. London, UK: Imperial College Press: pp. 15–32.

Alexy, O., Criscuolo, P., Salter, A. (2011). No soliciting: strategies for managing unsolicited innovative ideas. *California Management Review*, **54**, 116–139.

Allocca, M.A., Kessler, E.H. (2006). Innovation speed in small and medium-sized enterprises. *Creativity & Innovation Management*, **15**, 279–295.

Amabile, T. (1983). The social psychology of creativity: a componential conceptualization. *Journal of Personality and Social Psychology*, **45**, 357–376.

Amabile, T.M. (1996). *Creativity in Context: Update to the Social Psychology of Creativity.* Boulder, CO: Westview Press.

Amabile, T.M., Conti, R., Coon, H., Lazenby, J., Herron, M. (1996). Assessing the work environment for creativity. *Academy of Management Journal*, **39**, 1154–1184.

Amabile, T.M., Pillemer, J. (2012). Perspectives on the social psychology of creativity. *The Journal of Creative Behavior*, **46**, 3–15.

Anderson, J.R. (1983). A spreading activation theory of memory. *Journal of Verbal Learning and Verbal Behavior*, **22**, 261–295.

Antons, D., Piller, F.T. (2015). Opening the black box of 'not invented here': attitudes, decision biases, and behavioral consequences. *Academy of Management Perspectives*, **29**, 193–217.

Antorini, Y.M., Muñiz, A.M. (2013). The benefits and challenges of collaborating with user communities. *Research-Technology Management*, **56**, 21–28.

Arnould, E.J., Cayla, J. (2015). Consumer fetish: commercial ethnography and the sovereign consumer. *Organization Studies*, **36**, 1361–1386.

Athaide, G. A., Klink, R.R. (2009). Managing seller-buyer relationships during new product development. *Journal of Product Innovation Management*, **26**, 566–577.

Atuahene-Gima, K., Ko, A. (2001). An empirical investigation of the effect of market orientation and entrepreneurship orientation alignment on product innovation. *Organization Science*, **12**, 54–74.

Bagozzi, R.P., Dholakia, U.M. (2006). Open source software user communities: a study of participation in Linux user groups. *Management Science*, **52**, 1099–1115.

Baker, W.E., Sinkula, J.M. (2005). Market orientation and the new product paradox. *Journal of Product Innovation Management*, **22**, 483–502.

_____. (2007). Does market orientation facilitate balanced innovation programs? An organizational learning perspective. *Journal of Product Innovation Management*, **24**, 316–334.

Baldwin, C., Hienerth, C., von Hippel, E. (2006). How user innovations become commercial products: a theoretical investigation and case study. *Research Policy*, **35**, 1291–1313.

Bao, Y., Sheng, S., Zhou, K.Z. (2012). Network-based market knowledge and product innovativeness. *Marketing Letters*, **23**, 309–324.

Barlow, J.B., Giboney, J., Keith, M.J., Wilson, D., Schuetzler, R.M., Lowry, P.B., Vance, A. (2011). Overview and guidance on Agile development in large organizations. *Communications of the Association for Information Systems*, **29**, 25–44.

Barron, F. (1968). *Creativity and Personal Freedom.* New York, NY: van Nostrand.

Bartl, M., Füller, J., Mühlbacher, H., Ernst, H. (2012). A manager's perspective on virtual customer integration for new product development. *Journal of Product Innovation Management*, **29**, 1031–1046.

Barton, C., Koslow, L., Dhar, R., Chadwick, S., Reeves, M. (2016). Why companies can't turn customer insights into growth. Boston Consulting Group. Available at https://www.bcgperspectives.com/content/articles/center-customer-insight-marketing-sales-why-companies-cant-turn-customer-insights-growth/ (accessed 30 August 2017).

Baxter, D.I., Goffin, K., Szwejczewski, M. (2014). The repertory grid technique as a customer insight method. *Research Technology Management*, **57**, 35–42.

Bayus, B.L. (2013). Crowdsourcing new product ideas over time: an analysis of the dell ideastorm community. *Management Science*, **59**, 226–244.

Beckley, J., Paredes, D., Lopetcharats, K. (2012). *Product Innovation Toolbox: A Field Guide to Consumer Understanding and Research.* Oxford, UK: Wiley-Blackwell.

Belk, R. (2014). Digital consumption and the extended self. *Journal of Marketing Management*, **1376**, 1–18.

Belk, R., Fischer, E., Kozinets, R.V. (2013). *Qualitative Consumer and Marketing Research.* London, UK: Sage.

Belz, F.-M., Baumbach, W. (2010). Netnography as a method of lead user identification. *Creativity and Innovation Management*, **19**, 304–313.

Benaroch, M., Lichtenstein, Y., Robinson, K. (2006). Real options in information technology risk management: an empirical validation of risk-option relationships. *MIS Quarterly*, **30**, 827–864.

Benkler, Y. (2002). Coase's penguin, or, Linux and the nature of the firm. *Yale Law Journal*, **112**, 369–446.

Berchicci, L., Tucci, C.L. (2010). There is more to market learning than gathering good information: the role of shared team values in radical product definition. *Journal of Product Innovation Management*, **27**, 972–990.

Berekoven, L., Eckert, W., Ellenrieder, P. (2004). *Marktforschung: Methodische Grundlagen und praktische Anwendung.* 10th editition. Wiesbaden: Gabler.

Bessant, J., Maher, L. (2009). Developing radical service innovations: the role of design methods. *International Journal of Innovation Management*, **13**, 555–568.

Bilgram, V., Brem, A., Voigt, K.-I. (2008). User-centric innovations in new product development — systematic identification of lead users harnessing interactive and collaborative online-tools. *International Journal of Innovation Management*, **12**, 419–458.

Bilgram, V., Füller, J., Bartl, M., Biel, S., Miertsch, H. (2013). Eine Allianz gegen Flecken. *Harvard Business Manager*, **3**, 362–368.

Bitner, M.J., Booms, B.H., Tetreault, M.S. (1990). The service encounter: diagnosing favorable and unfavorable incidents. *The Journal of Marketing*, **54**, 71–84.

Bitner, M.J., Ostrom, A.L., Morgan, F.N. (2008). Service blueprinting: a practical technique for service innovation. *California Management Review*, **50**, 66–94.

Boddy, C. (2011). Hanging around with people: ethnography in marketing research and intelligence gathering. *The Marketing Review*, **11**, 151–163.

Bohlmann, J.D., Spanjol, J., Qualls, W.J., Rosa, J.A. (2013). The interplay of customer and product innovation dynamics: an exploratory study. *Journal of Product Innovation Management*, **30**, 228–244.

Bonner, J.M. (2005). The influence of formal controls on customer interactivity in new product development. *Industrial Marketing Management*, **34**, 63–69.

Boudreau, K., Lacetera, N., Lakhani, K. (2011). Looking across and looking beyond the knowledge frontier: intellectual distance, novelty, and resource allocation in science. *Management Science*, **57**, 843–863.

Brem, A., Voigt, K.-I. (2009). Integration of market pull and technology push in the corporate front end and innovation management—insights from the German software industry. *Technovation*, **29**, 351–367.

Brockhoff, K. (2003). Customer's perspectives of involvement in new product development. *International Journal of Technology Management*, **26**, 464–481.

Brown, J.S., Duguid, P. (2001). Knowledge and organization: a social-practice perspective. *Organization Science*, **12**, 198–213.

Buell, R.W., Kim, T., Tsay, C.-J. (2016). Creating reciprocal value through operational transparency. *Management Science*, **63**, 1673–1695.

Bullinger, A., Neyer, A., Rass, M., Moeslein, K. (2010). Community-based innovation contests: where competition meets cooperation. *Creativity and Innovation Management*, **19**, 290–303.

Burroughs, J., Moreau, C.P., Mick, D. (2008). Toward a psychology of consumer creativity. In: Haugtvedt, C.P., Herr, P.M., Kardes, F.R. (eds.) *Handbook of Consumer Psychology*. 38th Edition. New York, NY: Erlbaum: pp. 1011–1038.

Butzerin, D.J. (2011). Kundenintegration in den frühen phasen des innovationsprozesses am beispiel von Hiltis drehmomentkontrollierten anwendungen. Dissertation, University of Applied Sciences Upper Austria, Wels.

Cameron, J. (2001). Negative Effects of reward on intrinsic motivation: a limited phenomenon: comment on Deci, Koestner, and Ryan. *Review of Educational Research*, **71**, 29–42.

Carlson, R.N., Heth, C.D. (2007). *Psychology: The Science of Behaviour*. Upper Saddle River, NJ: Pearson Education.

Chandy, R.K., Tellis, G.J. (1998). Organizing for radical product innovation: the overlooked role of willingness to cannibalize. *Journal of Marketing Research*, **35**, 474–487.

Chase, W.G., Simon, H.A. (1973). Perception in chess. *Cognitive Psychology*, **4**, 55–61.

Chatterji, A.K., Fabrizio, K.R. (2013). Using users: when does external knowledge enhance corporate product innovation? *Strategic Management Journal*, **35**, 1427–1445.

Chell, E. (1998). *Critical Incident Technique in Qualitative Methods and Analysis in Organizational Research: A Practical Guide*. Thousand Oaks, CA: Sage: pp. 51–72.

Christensen, C.M., Bower, J.L. (1996). Customer power, strategic investment, and the failure of leading firms. *Strategic Management Journal*, **17**, 197–218.

Christensen, G.L., Olson, J.C. (2002). Mapping Consumers' mental models with ZMET. *Psychology & Marketing*, **19**, 477–501.

Churchill, J., von Hippel, E., Sonnack, M. (2009). *Lead User Project Handbook. A Practical Guide for Lead User Project Teams*. Cambridge, MA: MIT Press.

Coase, R.H. (1937). The Nature of the Firm. In: Williamson, O.E., Winter, S.G. (eds.) *The Nature of the Firm: Origins, Evolution, and Development*. New York, NY: Oxford University Press: pp. 18–33.

Cobb, C.G. (2011). *Making Sense of Agile Project Management: Balancing Control and Agility*. Hoboken, NJ: Wiley.

Cohen, M.W., Levinthal, D.A. (1990). Absoprtive capacity: a new perspective on learning and innovation. *Administrative Science Quarterly*, **35**, 128–152.

Collins, M.A., Amabile, T.M. (1999). Motivation and creativity. In Sternberg, R.J. (ed.) *Handbook of Creativity*. Cambridge, MA: Cambridge University Press: pp. 297–312.

Conforto, E.C., Amaral, D.C. (2015). Agile project management and stage-gate model — a hybrid framework for technology-based companies. *Journal of Engineering and Technology Management*, **40**, 1–14.

Cooper, R.G. (1990). Stage-gate systems: a new tool for managing new products. *Business Horizons*, **33**, 44–54.

_____. (1994). Third-generation new product processes. *Journal of Product Innovation Management*, **11**, 3–14.

_____. (2008). Perspective: the stage-gate idea-to-launch process — update, what's new, and NextGen systems. *Journal of Product Innovation Management*, **25**, 213–232.

Cooper, R.G., Edgett, S.J., Kleinschmidt, E.J. (2002). Optimizing the stage-gate process: what best-practice companies do. *Research Technology Management*, **45**, 21–27.

Cooper, R.G., Kleinschmidt, E.J. (1991). New product processes at leading industrial firms. *Industrial Marketing Management*, **20**, 137–147.

Cooper, R.G., Sommer, A.F. (2016). The agile-stage-gate hybrid model: a promising new approach and a new research opportunity. *Journal of Product Innovation Management*, **33**, 513–526.

Crawford, C.M., Di Benedetto, C.A. (2008). *New Products Management*. 9th Edition. New York, NY: McGraw-Hill Irwin.

Criscuolo, P., Dahlander, L., Grohsjean, T., Salter, A. (2017). Evaluating novelty: the role of panels in the selection of R&D projects. *Academy of Management Journal*, **60**, 433–460.

Crowe, E., Higgins, E.T. (1997). Regulatory focus and strategic inclinations: promotion and prevention in decision-making. *Organizational Behavior and Human Decision Processes*, **69**, 117–132.

Csikszentmihalyi, M. (1994). The domain of creativity. In: Feldman, D.H., Csikszentmihalyi, M., Gardnerk, H. (eds.) *Changing the World: A Framework for the Study of Creativity*. Westport, CT: Praeger Publishers: pp. 135–158.

_____. (1996). *Creativity Flow and the Psychology of Discovery and Invention*. New York, NY: Harper Perennial.

Cuny, C., Opaswongkarn, T. (2017). 'Why do young Thai women desire white skin?' Understanding conscious and nonconscious motivations of young women in Bangkok. *Psychology & Marketing*, **34**, 556–568.

Daetz, D., Barnard, B., Norman, R. (1995). *Customer Integration: The Quality Function Deployment (QFD) Leader's Guide for Decision Making*. New York, NY: Wiley.

Dahan, E., Kim, A.J., Lo, A.W., Poggio, T., Chan, N. (2011). Securities trading of concepts (STOC). *Journal of Marketing Research*, **48**, 497–517.

Dahan, E., Srinivasan, V. (2000). The predictive power of internet-based product concept testing using visual depiction and animation. *Journal of Product Innovation Management*, **17**, 99–109.

Dahl, D.W., Fuchs, C., Schreier, M. (2015). Why and when consumers prefer products of user-driven firms: a social identification account. *Management Science*, **61**, 1978–1988.

Dahlander, L., Piezunka, H. (2014). Open to suggestions: how organizations elicit suggestions through proactive and reactive attention. *Research Policy*, **43**, 812–827.

Danneels, E. (2003). Tight-loose coupling with customers: the enactment of customer orientation. *Strategic Management Journal*, **24**, 559–576.

Davis, M.H. (1980). A multidimensional approach to individual differences in empathy. *JSAS Catalog of Selected Documents in Psychology*, **10**, 85–102.

Davis, R.E. (1993). From experience: the role of market research in the development of new consumer products. *Journal of Product Innovation Management*, **10**, 309–317.

Day, G., Shoemaker, P.J.H. (2005). Scanning the periphery. *Harvard Business Review*, **83**, 135–148.

Denicolai, S., Ramirez, M., Tidd, J. (2016). Overcoming the false dichotomy between internal R&D and external knowledge acquisition: absorptive capacity dynamics over time. *Technological Forecasting and Social Change*, **104**, 57–65.

Desouza, K.C., Awazu, Y., Jha, S., Dombrowski, C., Papagari, S., Baloh, P., Kim, J.Y. (2008). Customer-driven innovation. *Research Technology Management*, **51**, 35–44.

Dholakia, U.M. (1997). An investigation of the relationship between perceived risk and product involvement. *Advances in Consumer Research*, **24**, 159–167.

Dickinson, J.R., Wilby, C.P. (1997). Concept testing with and without product trial. *Journal of Product Innovation Management*, **14**, 117–125.

Dolan, R. (1993). *Managing the New Product Process*. Reading, MA: Addison-Wesley Publishing Company.

Dolan, R.J., Matthews, J.M. (1993). Maximizing the utility of customer product testing: beta test design and management. *Journal of Product Innovation Management*, **10**, 318–330.

Domzal, T.J., Unger, L.S. (1985). Judgments of verbal versus pictorial presentations of a product with functional and aesthetic features. *Advances in Consumer Research*, **12**, 268–272.

van den Driest, F., Sthanunathan, S., Weed, K. (2016). Building an insights engine. *Harvard Business Review*, **94**, 64–74.

Droge, C., Stanko, M.A., Pollitte, W.A. (2010). Lead users and early adopters on the web: the role of new technology product blogs. *Journal of Product Innovation Management*, **27**, 66–82.

Duncker, K. (1945). On problem solving. *Psychological Monographs*, **58**, 1–110.
Durgee, J.F., O'Connor, G.C., Veryzer, J.R.W. (1998). Using mini-concepts to identify opportunities for really new product functions. *Journal of Consumer Marketing*, **15**, 525–543.
Duysters, G., Kok, G., Vaandrager, M. (1999). Crafting successful strategic technology partnerships. *R&D Management*, **29**, 343–351.
Easterby-Smith, M. (1980). The design, analysis and interpretation of repertory grids. *International Journal of Man-Machine Studies*, **13**, 3–24.
Ebner, W., Leimeister, J.M., Krcmar, H. (2009). Community engineering for innovations: the ideas competition as a method to nurture a virtual community for innovations. *R&D Management*, **39**, 342–356.
Eisenberg, I. (2011). Lead-user research for breakthrough innovation. *Research Technology Management*, **54**, 50–58.
Eisenberger, R., Cameron, J. (1996). Detrimental effects of reward: reality or myth? *American Psychologist*, **51**, 1153–1166.
Enkel, E., Bader, K. (2016). Why do experts contribute in cross-industry innovation? A structural model of motivational factors, intention, and behavior. *R&D Management*, **46**, 207–226.
Enkel, E., Kausch, C., Gassmann, O. (2005). Managing the risk of customer integration. *European Management Journal*, **23**, 203–213.
Ericsson, K.A., Lehmann, A.C. (1996). Expert and exceptional performance: evidence of maximal adaptation to task constraints. *Annual Review of Psychology*, **47**, 273–305.
Ettlie, J.E., Elsenbach, J.M. (2007). Modified stage-gate regimes in new product development. *Journal of Product Innovation Management*, **24**, 20–33.
Evans, S. (2010). Winner of Starbucks' coffee cup challenge isn't a coffee cup. Available at https://www.fastcompany.com/1661306/winner-starbucks-coffee-cup-challenge-isnt-coffee-cup (accessed 30 August 2017).
Fiol, C.M., Lyles, M. (1985). Organizational learning. *Academy of Management Review*, **10**, 803–813.
Fixson, S.K., Marion, T.J. (2016). When innovation stumbles: limits to open innovation? Paper presented at the Open and User Innovation Conference, 1–31 August, 2016, Glasgow.
Flanagan, J.C. (1954). The critical incident technique. *Psychological Bulletin*, **51**, 327–358.
Fleming, L., Sorenson, O. (2004). Science as a map in technological search. *Strategic Management Journal*, **25**, 909–928.
Fourt, L.A., Woodlock, J.W. (1960). Early prediction of market success for new grocery products. *Journal of Marketing*, **25**, 31–38.

Franke, N., von Hippel, E. (2003). Satisfying heterogeneous user needs via innovation toolkits: the case of apache security software. *Research Policy*, **32**, 1199–1215.

Franke, N., von Hippel, E., Schreier, M. (2006). Finding commercially attractive user innovations: a test of lead-user theory. *Journal of Product Innovation Management*, **23**, 301–315.

Franke, N., Schreier, M., Kaiser, U. (2010). The 'I designed it myself' effect in mass customization. *Management Science*, **56**, 125–140.

Franke, N., Shah, S. (2003). How communities support innovative activities: an exploration of assistance and sharing among end-users. *Research Policy*, **32**, 157–178.

Frensch, P.A., Sternberg, R.J. (1989). Expertise and intelligent thinking: when is it worse to know better? In: Sternberg, R. (ed.) *Advances in Psychology of Human Intelligence*, Volume 5. Hillsdale, NJ: Erlbaum: pp. 157–188.

Friedman, M., Schillewaert, N. (2012). Order and quality effects in sequential monadic concept testing: methodological details matter in concept-testing practice. *Journal of Marketing Theory & Practice*, **20**, 377–390.

Fu, J., Matzler, K., Hoppe, M. (2008). Brand community members as a source of innovation. *Journal of Product Innovation Management*, **25**, 608–619.

Fuchs, C., Prandelli, E., Schreier, M., Dahl, D.W. (2013). All that is users might not be gold: how labeling products as user designed backfires in the context of luxury fashion brands. *Journal of Marketing*, **77**, 75–91.

Fuchs, C., Schreier, M. (2011). Customer empowerment in new product development. *Journal of Product Innovation Management*, **28**, 17–32.

Füller, J. (2010). Refining virtual co-creation from a consumer perspective. *California Management Review*, **52**, 98–122.

Füller, J., Hutter, K., Faullant, R. (2011). Why co-creation experience matters? Creative experience and its impact on the quantity and quality of creative contributions. *R&D Management*, **41**, 259–274.

Füller, J., Jawecki, G., Mühlbacher, H. (2007). Innovation creation by online basketball communities. *Journal of Business Research*, **60**, 60–71.

Füller, J., Matzler, K. (2007). Virtual product experience and customer participation: a chance for customer-centred, really new products. *Technovation*, **27**, 378–387.

Füller, J., Matzler, K., Hoppe, M. (2008). Brand community members as a source of innovation. *Journal of Product Innovation Management*, **25**, 608–619.

Füller, J., Matzler, K., Hutter, K., Hautz, J. (2012). Consumers' creative talent: which characteristics qualify consumers for open innovation projects? An exploration of asymmetrical effects. *Creativity and Innovation Management*, **21**, 247–262.

Di Gangi, P., Wasko, M., Hooker, R. (2010). Getting customers' ideas to work for you: learning from dell how to succeed with online user innovation communities. *MIS Quarterly Executive*, **9**, 213–228.

Garcia, R., Bardhi, F., Friedrich, C. (2007). Overcoming consumer resistance to innovation. *MIT Sloan Management Review*, **48**, 82–88.

Gassmann, O. (2013). *Crowdsourcing: Innovationsmanagement mit Schwarmintelligenz*. Munich: Carl Hanser.

Gassmann, O., Kausch, C., Enkel, E. (2010). Negative side effects of customer integration. *International Journal of Technology Management*, **50**, 43–62.

Gassmann, O., Schweitzer, F. (2014). *Management of the Fuzzy Front End of Innovation*. Berlin: Springer.

Gaubinger, K., Rabl, M., Swan, S., Werani, T. (2015). *Innovation and Product Management: A Holistic and Practical Approach to Uncertainty Reduction*. Berlin: Springer.

Gebauer, H., Worch, H., Truffer, B. (2012). Absorptive capacity, learning processes and combinative capabilities as determinants of strategic innovation. *European Management Journal*, **30**, 57–73.

Geiger, D., Seedorf, S., Schulze, T., Nickerson, R.C., Schader, M. (2011). Managing the crowd : towards a taxonomy of crowdsourcing processes. In: Proceedings of the Seventeenth Americas Conference on Information Systems (AMCIS), Detroit, Michigan, 4–7 August, 2011.

Gerhart, B., Rynes, S.L., Fulmer, I.S. (2009). Pay and performance: individuals, groups, and executives. *The Academy of Management Annals*, **3**, 207–271.

Geschka, H. (1986). Creativity workshops in product innovation. *Journal of Product Innovation Management*, **3**, 48–56.

Getzels, J.W. (1987). Creativity, intelligence, and problem finding: retrospect and prospect. In: *Frontiers of Creativity Research: Beyond the Basics*. Buffalo, NY: Bearly: pp. 88–102.

Gilleran, R., Noyes, E. (2015). *MakerBot: Challenges in Building a New Industry*. Wellesley, MA: Babson College: pp. 1–16.

Glaser, B.G., Strauss, A.L. (1967). *The Discovery of Grounded Theory: Strategies for Qualitative Research*. Chicago: Aldine Publishing Company.

Goffin, K., Lemke, F., Koners, U. (2010). *Identifying Hidden Needs*. New York, NY: Palgrave Macmillan.

Govindarajan, V., Kopalle, P.K. (2006). Disruptiveness of innovations: measurement and an assessment of reliability and validity. *Strategic Management Journal*, **27**, 189–199.

Govindarajan, V., Kopalle, P.K., Danneels, E. (2011). The effects of mainstream and emerging customer orientations on radical and disruptive innovations. *Journal of Product Innovation Management*, **28**, 121–132.

Green, P.E., Srinivasan, V. (1990). Conjoint analysis in marketing: new development with implications for research and practice. *Journal of Marketing*, **54**, 3–19.

Greifeneder, R., Bless, H., Pham, M.T. (2011). When do people rely on affective and cognitive feelings in judgment? A review. *Personality and Social Psychology Review*, **15**, 107–41.

Gremler, D.D. (2004). The critical incident technique in service research. *Journal of Service Research*, **7**, 65–89.

Griffin, A. (1997). PDMA research on new product development practices: Updating trends and benchmarking best practices. *Journal of Product Innovation Management*, **14**, 429–458.

Griffin, A., Price, R.L., Maloney, M.M., Vojak, B.A., Sim, E.W. (2009). Voices from the field: How exceptional electronic industrial innovators innovate. *Journal of Product Innovation Management*, **26**, 222–240.

Guilford, J.P. (1967). *The Nature of Human Intelligence*. New York, NY: McGraw-Hill.

Gulati, R., Oldroyd, J.B. (2005). The quest for customer focus. *Harvard Business Review*, **83**, 92–101.

Gupta, A.K., Wilemon, D.L. (1990). Accelerating the development of technology-based new products. *California Management Review*, **32**, 24–44.

Gustafsson, A., Herrmann, A., Huber, F. (2007). *Conjoint Measurement: Methods and Applications*. 4th Edition. Berlin: Springer.

Gutman, J. (1982). A means-end chain model based on consumer categorization processes. *Journal of Marketing*, **46**, 60–72.

Hagstotz, W., Schmitt-Hagstotz, K. (2008). Car clinic. In: Pepels, W. (ed.) *Marktforschung — Organisation und praktische Anwendung*. Düsseldorf: Symposion Publishing: pp. 513–531.

Haiken, M. (2016). Inspect and Adapt: Implementing Agile in GE's Global Research Environment. *Agilvox — Speak Your Mindset*, **3**, 6–15.

Hales, M., Tidd, J. (2009). Routines and representations at work in new product development. *Industrial and Corporate Change*, **18**, 551–574.

Hallen, B.L., Katila, R., Rosenberger, J.D. (2014). How do social defenses work? A resource-dependence lens on technology ventures, venture captial investors, and corporate relationships. *Academy of Management Journal*, **57**, 1078–1101.

Hambrick, D.C., Fredrickson, J.W. (2005). Are you sure you have a strategy? *Academy of Management Executive*, **19**, 51–62.

Hamel, G., Prahalad, C.K. (1994). *Competing for the Future: Breakthrough Strategies for Seizing Control of Your Industry and Creating the Markets for Tomorrow.* Cambridge, MA: Harvard Business School Press.

Hammarfjord, M.O., Roxenhall, T. (2017). The relationships between network commitment, antecedents, and innovation in strategic innovation networks. *International Journal of Innovation Management*, **21**, 1750037.

Hammond, K.R. (1996). *Human Judgment and Social Policy: Irreducible Uncertainty, Inevitable Error, Unavoidable Injustice*. New York, NY: Oxford University Press.

Han, J.K., Kim, N., Srivastava, R.K. (1998). Market orientation and organizational performance : Is innovation a missing link? *Journal of Marketing*, **62**, 30–45.

Harhoff, D., Henkel, J., von Hippel, E. (2003). Profiting from voluntary information spillovers: how users benefit by freely revealing their innovations. *Research Policy*, **32**, 1753–1769.

Harmancioglu, N., McNally, C.R., Calantone, R.J., Durmusoglu, S. (2007). Your new product development (NPD) is only as good as your process. *R&D Management*, **37**, 399–424.

Hattula, J.D., Herzog, W., Dahl, D.W., Reinecke, S. (2015). Managerial empathy facilitates egocentric predictions of consumer preferences. *Journal of Marketing Research*, **52**, 235–252.

Hemetsberger, A., Godula, G. (2007). Virtual customer integration in new product development in industrial markets: the QLL framework. *Journal of Business-to-Business Marketing*, **14**, 1–37.

van den Hende, E.A., Dahl, D.W., Schoormans, J.P.L., Snelders, D. (2012). Narrative transportation in concept tests for really new products: the moderating effect of reader-protagonist similarity. *Journal of Product Innovation Management*, **29**, 157–170.

Henderson, R., Clark, K.B. (1990). Architectural innovation: the reconfiguration of existing product technologies and the failure of established firms. *Administrative Sciences Quarterly*, **35**, 9–30.

Hertel, G., Niedner, S., Herrmann, S. (2003). Motivation of software developers in Open Source projects: an Internet-based survey of contributors to the Linux kernel. *Research Policy*, **32**, 1159–1177.

Hewing, M. (2013). Merits of collaboration with potential and current users in creative problem-solving. *International Journal of Innovation Management*, **17**, 1–27.

Hienerth, C. (2006). The commercialization of user innovations: the development of the rodeo kayak industry. *R&D Management*, **36**, 273–294.

_____. (2007). Exploring key characteristics of lead user workshop participants: who contributes best to the generation of truly novel solutions? Proceedings of the DRUID Summer Conference 2007 on Appropriability, Proximity, Routines and Innovation, Copenhagen, Denmark.

Hienerth, C., Lettl, C. (2011). Exploring how peer communities enable lead user innovations to become standard equipment in the industry: community pull effects. *Journal of Product Innovation Management*, **28**, 175–195.

Higgins, E.T. (1997). Beyond pleasure and pain. *American Psychologist*, **52**, 1280–1300.

_____. (1998). Promotion and prevention: regulatory focus as a motivational principle. *Advances in Experimental Social Psychology*, **30**, 1–46.

Hill, C.W.L., Jones, T.M. (1992). Stakeholder-agency theory. *Journal of Management Studies*, **29**, 131–154.

von Hippel, E. (1986). Lead users: a source of novel product concepts. *Management Science*, **32**, 791–805.

_____. (1994). "Sticky Information" and the locus of problem solving: implications for innovation. *Management Science*, **40**, 429–439.

_____. (2005). *Democratizing Innovation*. Cambridge, MA: MIT Press.

von Hippel, E., Franke, N., Prügl, R. (2009). Pyramiding: efficient search for rare subjects. *Research Policy*, **38**, 1397–1406.

von Hippel, E., Katz, R. (2002). Shifting innovation to users via toolkits. *Management Science*, **48**, 821–833.

von Hippel, E., von Krogh, G. (2017). CROSSROADS—identifying viable "need–solution pairs": problem solving without problem formulation. *Organization Science*, **27**, 207–221.

von Hippel, E., Thomke, S., Sonnack, M. (2000). Creating breakthroughs at 3M. *Harvard Business Review*, **77**, 47–57.

Hoeffler, S. (2003). Measuring preferences for really new products. *Journal of Marketing Research*, **40**, 406–420.

Hoffman, D.L., Kopalle, P.K., Novak, T.P. (2010). The "right" consumers for better concepts: identifying consumers high in emergent nature to develop new product concepts. *Journal of Marketing Research*, **47**, 854–865.

Holahan, P.J., Sullivan, Z.Z., Markham, S.K. (2014). Product development as core competence: how formal product development practices differ for radical, more innovative, and incremental product innovations. *Journal of Product Innovation Management*, **31**, 329–345.

Holt, K., Geschka, H., Peterlongo, G. (1984). *Need Assessment, a Key to User-Oriented Product Innovation*. Chichester, UK: Wiley.

Homburg, C., Workman, J.P., Krohmer, H. (1999). Marketing's influence within the firm. *Journal of Marketing*, **63**, 1–17.

Hoyer, W.D., Chandy, R.K., Dorotic, M., Krafft, M., Singh, S.S. (2010). Consumer cocreation in new product development. *Journal of Service Research*, **13**, 283–296.

Huber, J. (2005). Conjoint analysis: how we got here and where we are (an update). Research Paper Series. Sequim, WA: Sawtooth Software, Inc., pp. 1–15.

Hull, F.M., Storey, C. (2016). *Total Value Development: How to Drive Service Innovation*. London, UK: Imperial College Press.

Innocentive. (2015). Available at https://www.innocentive.com/about-us/ (accessed 30 August 2017).

Isaksen, S.G., Dorval, K.B., Treffinger, D.J. (1994). *Creative Approaches to Problem Solving*. Dubuque, IA: Kendall Hunt.

Jägel, T., Keeling, K., Reppel, A., Gruber, T. (2012). Individual values and motivational complexities in ethical clothing consumption: a means-end approach. *Journal of Marketing Management*, **28**, 373–396.

Jankowicz, D. (2004). *The Easy Guide to Repertory Grids*. Chichester, UK: Wiley.

Jaworski, B., Kohli, A.K., Sahay, A. (2000). Market-driven versus driving markets. *Journal of the Academy of Marketing Science*, **28**, 45–54.

Jeppesen, L.B., Lakhani, K.R. (2013). Marginality and problem-solving effectiveness in broadcast search. *Organization Science*, **21**, 1016–1033.

Jiang, Z., Scheibe, K.P., Nilakanta, S., (Shawn) Qu, X. (2017). The economics of public beta testing. *Decision Sciences*, **48**, 150–175.

Johansen, B., Ronn, K. (2014). *The Reciprocity Advantage: A New Way to Partner for Innovation and Growth.* San Francisco, CA: Berrett-Koehler Publishers.

Johnson, P., Robinson, P. (2014). Civic hackathons: innovation, procurement, or civic engagement? *Review of Policy Research*, **31**, 349–357.

Johnson, S.P., Menor, L.J., Roth, Aleda, V., Chase, R.B. (2000). A critical evaluation of the new service development process: integrating service innovation and service design In: Fitzsimmons, J.A., Fitzsimmons, M.J. (eds.) *New Service Development: Creating Memorable Experiences*. London, UK: Sage: pp. 1–32.

Jungk, R., Müller, N. (1987). *Future Workshops: How to Create Desirable Futures*. London: Institute for Social Inventions.

Kahn, K.B. (2006). *New Product Forecasting: An Applied Approach.* New York, NY: M.E. Sharpe.

Kahneman, D. (1994). New challenges to the rationality assumption. *Journal of Institutional and Theoretical Economics*, **150**, 18–36.

Kaiserswerth, M. (2013). Building a bridge from research to the market: IBM's industry solutions labs. In: Gassmann, O., Schweitzer, F. (eds.) *Management of the Fuzzy Front End of Innovation*. Cham, Switzerland: Springer: pp. 283–288.

Kalbach, J. (2016). *Mapping Experiences: A Guide to Creating Value through Journeys, Blueprints, and Diagrams*. Sebastopol, CA: O'Reilly.

Kantar TNS. (2017). Konsumenten-gestützte Innovations-forschung. Available at https://www.tns-infratest.com/Branchen-und-Maerkte/pharma-innovations-forschung.asp (accessed 30 August 2017).

Kark, R., Dijk, D. (2007). Motivation to lead, motivation to follow: the role of the self-regulatory focus in leadership processes. *Academy of Management Review*, **32**, 500–528.

Katila, R., Ahuja, G. (2002). Something old, something new: a longitudinal study of search behavior and new product introduction. *Academy of Management Journal*, **45**, 1183–1194.

Katz, E., Blumler, J.G., Gurevitch, M. (1974). Utilization of mass communication by the individual. In: Blumler, J.G., Katz, E. (eds.) *The Use of Mass Communications: Current Perspectives on Gratifications Research*. Beverly Hills, CA: Sage: pp. 19–32.

Katz, R., Allen, T.J. (1988). Investigating the Not Invented Here (NIH) syndrome. In: Katz, R. (ed.) *Managing Professionals in Innovative Organizations*. Cambridge, MA: Ballinge: pp. 442–456.

Kelly, G.A. (1955). *A Theory of Personality*. New York, NY: W.W. Norton.

Kelly, T. (2002). *The Art of Innovation: Lessons in Creativity from IDEO*. New York, NY: HarperCollins Business.

Kirca, A.H., Jayachandran, S., Bearden, W.O. (2005). Market orientation: a meta-analytic review and assessment of its antecedents and impact on performance. *Journal of Marketing*, **69**, 24–41.

Kitchell, S. (1995). Corporate Culture, environmental adaptation, and innovation adoption: a qualitative/quantitative approach. *Journal of the Academy of Marketing Science*, **23**, 195–205.

Klink, R.R., Athaide, G.A. (2006). An illustration of potential sources of concept-test error. *Journal of Product Innovation Management*, **23**, 359–370.

Knudsen, M.P. (2007). The relative importance of interfirm relationships and knowledge transfer for new product development success. *Journal of Product Innovation Management*, **24**, 117–138.

Kok, R.A.W., Hillebrand, B., Biemans, W.G. (2003). What makes product development market oriented? Towards a conceptual framework. *International Journal of Innovation Management*, **7**, 137–162.

Kokshagina, O., Le Masson, P., Bories, F. (2017). Fast-connecting search practices: on the role of open innovation intermediary to accelerate the absorptive capacity. *Technological Forecasting and Social Change*, **120**, 232–239.

Kornish, L.J., Ulrich, K. (2014). The importance of the raw idea in innovation: testing the sow's ear hypothesis. *Journal of Marketing Research*, **LI**, 14–26.

Kosonen, M., Gan, C., Vanhala, M., Blomqvist, K. (2014). User motivation and knowledge sharing in idea crowdsourcing. *International Journal of Innovation Management*, **18**, 1–23.

Kotler, P., Armstrong, G. (2009). *Principles of Marketing*. 13th Edition. UK: Prentice Hall.

Kotler, P., Keller, K.L. (2006). *Marketing Management*. Uppe Saddle River, NJ: Prentice Hall.

Kozinets, R.V. (2002). The field behind the screen: using netnography for marketing research in online communities. *Journal of Marketing Research*, **39**, 61–72.

_____. (2006). Netnography 2.0. In: Belk, R.W. (ed) *Handbook of Qualitative Research Methods in Marketing*. Cheltenham: Edward Elgar Publishing Limited: pp. 129–142.

Kratzer, J., Lettl, C. (2009). Distinctive roles of lead users and opnion leaders in the social networks of schoolchildren. *Journal of Consumer Research*, **36**, 646–659.

Kristensson, P., Gustafsson, A., Archer, T. (2004). Harnessing the creative potential among users. *Journal of Product Innovation Management*, **21**, 4–14.

Kristensson, P., Magnusson, P.R. (2010). Tuning users' innovativeness during ideation. *Creativity and Innovation Management*, **19**, 147–159.

Kristensson, P., Magnusson, P.R., Matthing, J. (2002). Users as a hidden resource for creativity: findings from an experimental study on user involvement. *Creativity and Innovation Management*, **11**, 55–61.

von Krogh, G., von Hippel, E. (2006). The promise of research on open source software. *Management Science*, **52**, 975–983.

von Krogh, G., Spaeth, S., Lakhani, K.R. (2003). Community, joining, and specialization in open source software innovation: a case study. *Research Policy*, **32**, 1217–1241.

Kuisma, T., Laukkanen, T., Hiltunen, M. (2007). Mapping the reasons for resistance to Internet banking: a means-end approach. *International Journal of Information Management*, **27**, 75–85.

Kyriakou, H., Nickerson, J.V., Sabnis, G. (2017). Knowledge reuse for customization: metamodels in an open design community for 3D printing. *MIS Quarterly*, **41**, 315–334.

Lakhani, K., Jeppesen, L.B. (2007). Getting unusual suspects to solve R&D puzzles. *Harvard Business Review*, **85**, 30–32.

Lakhani, K.R. (2009). "InnoCentive.com (A)". Harvard Business School Case 608–170: pp. 1–22.

Lakhani, K.R., von Hippel, E. (2003). How open source software works: "free" user-to-user assistance. *Research Policy*, **32**, 923–943.

Landau, C., Leih, S., Russo, P., Teece, D. (2016). Wenig weisheit bei den vielen. *Harvard Business Manager*, **38**, 72–78.

Lane, P.J., Koka, B.R., Pathak, S. (2006). The reification of absorptive capacity: a critical review and rejuvenation of the construct. *Academy of Management Review*, **31**, 833–863.

Laursen, K., Salter, A. (2006). Open for innovation: The role of openness in explaining innovation performance among U.K. manufacturing firms. *Strategic Management Journal*, **27**, 131–150.

Lee, Y.-N., Walsh, J.P. (2016). Inventing while you work: knowledge, non-R&D learning and innovation. *Research Policy*, **45**, 345–359.

Lee, Y.-N., Walsh, J.P., Wang, J. (2015). Creativity in scientific teams: unpacking novelty and impact. *Research Policy*, **44**, 684–697.

Lees, G., Wright, M.A. (2004). The effect of concept formulation on concept test scores. *Journal of Product Innovation Management*, **21**, 389–400.

Lemke, F., Clark, M., Wilson, H. (2011). Customer experience quality: an exploration in business and consumer contexts using repertory grid technique. *Journal of the Academy of Marketing Science*, **39**, 846–869.

Le Masson, P., Magnusson, P. (2002). Towards an understanding of user contribution to the design of mobile telecommunication services. 9th International Product Development Conference, Nice, May 25 & 26.

Lenfle, S., Loch, C. (2010). Lost roots: how project management came to emphasize control over flexibility and novelty. *California Management Review*, **53**, 32–55.

Leonard-Barton, D. (1992). Core capabilities and core rigidities: a paradox in managing new product development. *Strategic Management Journal*, **13**, 111–126.

Leonard, D.A., Rayport, J. (1997). Spark innovation through empathic design. *Harvard Business Review*, **75**, 102–113.

Leonard, N.H., Beauvais, L.L., Scholl, R.W. (2005). A multi-level model of group cognitive style in strategic decision making. *Journal of Managerial Issues*, **17**, 119–138.

Lettl, C. (2007). User involvement competence for radical innovation. *Journal of Engineering and Technology Management*, **24**, 53–75.

Lettl, C., Herstatt, C., Gemuenden, H.G. (2006a). Learning from users for radical innovation. *International Journal of Technology Management*, **33**, 25–45.

Lettl, C., Herstatt, C., Gemuenden, H.G. (2006b). Users' contributions to radical innovation: evidence from four cases in the field of medical equipment technology. *R and D Management*, **36**, 251–272.

Lettl, C., Hienerth, C., Gemuenden, H.G. (2008). Exploring how lead users develop radical innovation: opportunity recognition and exploitation in the field of medical equipment technology. *IEEE Transactions on Engineering Management*, **55**, 219–233.

Levinthal, D.A., March, J.G. (1993). The myopia of learning. *Strategic Management Journal*, **14**, 95–112.

Lieberman, M.D. (2000). Intuition: a social cognitive neuroscience approach. *Psychological Bulletin*, **126**, 109–137.

Lilien, G.L., Morrison, P.D., Searls, K., Sonnack, M., von Hippel, E. (2002). Performance assessment of the lead user idea-generation process for new product development. *Management Science*, **48**, 1042–1059.

Lindgren, L.M., O'Connor, G.C. (2011). The role of future-market focus in the early stages of NPD across varying levels of innovativeness. *Journal of Product Innovation Management*, **28**, 787–800.

Litchfield, R.C. (2008). Brainstorming reconsidered: a goal-based view. *Academy of Management Review*, **33**, 649–688.

Littler, D., Leverick, F., Bruce, M. (1995). Factors affecting the process of collaborative product development: a study of UK manufacturers of information and communication technology products. *Journal of Product Innovation Management*, **12**, 16–32.

Lubart, T.I. (1994). Creativity. In: Sternberg, R.J. (ed.) *Thinking and Problem Solving: Handbook of Perception and Cognition*. 2nd Edition. New York, NY: Academic Press: pp. 289–332.

Lüthje, C. (2004). Characteristics of innovating users in a consumer goods field: an empirical study of sport-related product consumers. *Technovation*, **24**, 683–695.

Lüthje, C., Herstatt, C., von Hippel, E. (2005). User-innovators and 'local' information: the case of mountain biking. *Research Policy*, **34**, 951–965.

Macdonald, E.K., Kleinaltenkamp, M., Wilson, H.N. (2016). How business customers judge solutions: solution quality and value in use. *Journal of Marketing*, **80**, 96–120.

MacDonald, S. (1995). Too close for comfort? The strategic implications of getting close to the customer. *California Management Review*, **37**, 8–27.

Mack, T., Landau, C. (2015). Winners, losers, and deniers: self-selection in crowd innovation contests and the roles of motivation, creativity, and skills. *Journal of Engineering & Technology Management*, **37**, 52–64.

MacKinnon, D.W. (1961). *The Creative Person*. Berkley, CA: University of California Press.

Maes, J., Sels, L. (2014). SMEs' radical product innovation: the role of internally and externally oriented knowledge capabilities. *Journal of Small Business Management*, **52**, 141–163.

Magnusson, P.R. (2009). Exploring the contributions of involving ordinary users in ideation of technology-based services. *Journal of Product Innovation Management*, **26**, 578–593.

Magnusson, P.R., Kristensson, P., Hipp, C. (2010). Exploring the ideation patterns of ordinary users: the case of mobile telecommunications services. *International Journal of Product Development*, **11**, 289–309.

Mahajan, V., Muller, E., Wind, Y. (2000). *New-Product Diffusion Models*. Dordrecht: Kluwer Academic Publishers Group.

Malone, T.W., Yates, J., Benjamin, R. (1987). Electronic markets and electronic hierarchies. *Communications of the ACM*, **30**, 484–497.

March, J.G. (1991). Exploration and exploitation in organizational learning. *Organization Science*, **2**, 71–87.

Marks, M., Hoyt, D. (2013). Quirky: a business based on making invention accessible. Case No. GS84, Case Studies, Stanford Graduate School of Business, 2013.

McDermott, C.M., O'Connor, G.C. (2002). Managing radical innovation: an overview of emergent strategic issues. *Journal of Product Innovation Management*, **19**, 424–438.

McDonagh-Philp, D., Lebbon, C. (2015). The emotional domain in product design. *The Design Journal*, **3**, 31–43.

Meyer, M.H., Crane, F.G., Lee, C. (2016). Connecting ethnography to the business of innovation. *Business Horizons*, **59**, 699–711.

Michalko, M. (2006). *Thinkertoys: A Handbook of Creative-Thinking Techniques*. 2nd Edition. Berkeley, CA: Ten Speed Press.

Miron-Spektor, E., Erez, M., Naveh, E., Miron-Spektor, E., Erez, M., Naveh, E. (2011). The effect of conformist and attentive-to-detail members on team innovation: reconciling the innovation paradox. *Academy of Management Journal*, **54**, 740–760.

Moore, G.A. (1999). *Crossing the Chasm: Marketing and Selling High-Tech Products to Mainstream Customers.* New York, NY: HarperCollins.

Moore, W.L. (1982). Concept testing. *Journal of Business Research*, **10**, 279–294.

Moreau, C.P.P., Lehmann, D.R., Markman, A.B. (2001). Entrenched knowledge structures and consumer response to new products. *Journal of Marketing Research*, **38**, 14–29.

Morrison, P.D., Roberts, J.H., von Hippel, E. (2000). Determinants of user innovation and innovation sharing in a local market. *Management Science*, **46**, 1513–1528.

Morrison, P.D., Roberts, J.H., Midgley, D.F. (2004). The nature of lead users and measurement of leading edge status. *Research Policy*, **33**, 351–362.

Mugge, R., Dahl, D.W. (2013). Seeking the ideal level of design newness: consumer response to radical and incremental product design. *Journal of Product Innovation Management*, **30**, 34–47.

Mühlbacher, H., Füller, J., Huber, L. (2011). Online forum discussion-based forecasting of new product market performance. *Marketing ZFP*, **33**, 221–234.

Mullen, B., Johnson, C., Salas, E. (1991). Productivity loss in brainstorming groups: a meta-analytic integration. *Basic and Applied Social Psychology*, **12**, 3–23.

Mullins, J.W., Sutherland, D.J. (1998). New product development in rapidly changing markets: an exploratory study. *Journal of Product Innovation Management*, **15**, 224–236.

Muylle, S., Basu, A., Standaert, W., Decraene, W., Debrabander, F., Praet, S. (2016). How can B2B organizations engage consumers through the use of digital technology? Research report. Accenture Belux–Vlerick Digital 20/20 Chair. Brussels: Vlerick Business School.

Nambisan, S. (2002). Designing virtual customer environments for new product development: toward a theory. *Academy of Management Review*, **27**, 392–413.

Nambisan, S., Baron, R.A. (2009). Virtual customer environments: testing a model of voluntary participation in value co-creation activities. *Journal of Product Innovation Management*, **26**, 388–406.

Narver, J., Slater, S. (1990). The effect of a market orientation on business profitability. *Journal of Marketing*, **54**, 20–35.

Narver, J.C., Slater, S.F., MacLachlan, D.L. (2004). Responsive and proactive market orientation and new-product success. *Journal of Product Innovation Management*, **21**, 334–347.

Natalicchio, A., Petruzzelli, A.M., Garavelli, A.C. (2014). A literature review on markets for ideas: emerging characteristics and unanswered questions. *Technovation*, **34**, 65–76.

Netzer, O., Srinivasan, V. (2011). Adaptive self-explication of multiattribute preferences. *Journal of Marketing Research*, **48**, 140–156.

Nijssen, E.J., Hillebrand, B., de Jong, J.P.J., Kemp, R.G.M. (2012). Strategic value assessment and explorative learning opportunities with customers. *Journal of Product Innovation Management*, **29**, 91–102.

Nishikawa, H., Schreier, M., Ogawa, S. (2013). User-generated versus designer-generated products: a performance assessment at Muji. *International Journal of Research in Marketing*, **30**, 160–167.

O'Connor, G.C., Rice, M.P. (2013). New market creation for breakthrough innovations: enabling and constraining mechanisms. *Journal of Product Innovation Management*, **30**, 209–227.

Obrist, M. (2007). Finding individuality in the technological complexity: why people do it themselves. *International Journal of Interdisciplinary Social Sciences*, **2**, 203–211.

Olson, E.L., Bakke, G. (2001). Implementing the lead user method in a high technology firm: a longitudinal study of intentions versus actions. *Journal of Product Innovation Management*, **18**, 388–395.

Orme, B.K. (2013). *Getting Started with Conjoint Analysis*. 3rd Edition. Madison, WI: Research Publishers LCC.

Østergaard, C.R., Timmermansa, B., Kristinsson, K. (2011). Does a different view create something new? The effect of employee diversity on innovation. *Research Policy*, **40**, 500–509.

Page, A.L., Rosenbaum, H.F. (1992). Developing an effective concept-testing program for consumer durables. *Journal of Product Innovation Management*, **9**, 267–277.

Parry, M.E., Song, M. (2010). Market information acquisition, use, and new venture performance. *Journal of Product Innovation Management*, **27**, 1112–1126.

Patton, M.Q. (2002). *Qualitative Research and Evaluation Methods*. 3rd Edition. Thousand Oaks, CA: Sage.

Peng, L., Finn, A. (2008). Concept testing: the state of contemporary practice. *Marketing Intelligence & Planning*, **26**, 649–674.

Peng, L., Finn, A. (2010). Whose crystal ball to choose? Individual difference in the generalizability of concept testing. *Journal of Product Innovation Management*, **27**, 690–704.

Pera, R., Occhiocupo, N., Clarke, J. (2016). Motives and resources for value co-creation in a multi-stakeholder ecosystem: a managerial perspective. *Journal of Business Research*, **69**, 4033–4041.

Pfeffer, J., Novak, P. (1978). *The External Control of Organizations: A Resource Dependence Perspective*. New York, NY: Harper & Row.

Picot, A., Rechwald, R., Wigand, R. (2003). *Die grenzenlose Unternehmung*. 5th Edition. Wiesbaden: Gabler.

Pierce, J.L., Delbecq, A.L. (1977). Organizational structure, individual attitudes, and innovation. *Academy of Management Review*, **2**, 26–37.

Pieters, R., Baumgartner, H., Allen, D. (1995). A means-end chain approach to consumer goal structures. *International Journal of Research in Marketing*, **12**, 227–244.

Piezunka, H., Dahlander, L. (2015). Distant search, narrow attention: how crowding alters organizations' filtering of suggestions in crowdsourcing. *Academy of Management Journal*, **58**, 856–880.

Piller, F., Ihl, C. (2009). Open Innovation with Customers Foundations, Competences and International Trends. Report. Aachen, Germany: Technology and Innovation Management Group at RWTH Aachen University (RWTH-TIM): pp. 1–67.

Piller, F.T., Walcher, D. (2006). Toolkits for idea competitions: a novel method to integrate users in new product development. *R&D Management*, **36**, 307–318.

Pine, J., Gilmore, J.H. (1999). *The Experience Economy*. Boston, MA: Harvard Business School Press.

Pisano, G.P., Verganti, R. (2008). Which kind of collaboration is right for you? *Harvard Business Review*, **86**, 79–86.

Poetz, M.K., Prügl, R.J. (2010). Crossing domain-specific boundaries in search of innovation: exploring the potential of pyramiding. *Journal of Product Innovation Management*, **27**, 897–914.

Poetz, M.K., Schreier, M. (2012). The value of crowdsourcing: can users really compete with professionals in generating new product ideas? *Journal of Product Innovation Management*, **29**, 245–256.

Porter, C.E., Donthu, N., MacElroy, W.H., Wydra, D. (2011). How to foster and sustain engagement in virtual communities. *California Management Review*, **53**, 80–110.

Poskela, J., Martinsuo, M. (2009). Management control and strategic renewal in the front end of innovation. *Journal of Product Innovation Management*, **26**, 671–684.

Postma, C.E., Zwartkruis-Pelgrim, E., Daemen, E., Du, J. (2012). Challenges of doing empathic design: experiences from industry. *International Journal of Design*, **6**, 59–70.

Prahalad, C.K., Ramaswamy, V. (2004). Co-creating unique value with customers. *Strategy & Leadership*, **32**, 4–9.

Prandelli, E., Verona, G., Raccagni, D. (2006). Diffusion of web-based product innovation. *California Management Review*, **48**, 109–135.

Prielinger, B. (2011). Instrumente zur einbindung von kundenbedürfnissen in die produktentwicklung internationaler industriegüterunternehmen. Dissertation, University of Applied Sciences Upper Austria, Wels.

Proctor, T. (2013). *Creative Problem Solving for Managers: Developing Skills for Decision Making and Innovation*. 4th Edition. New York, NY: Routledge Chapman & Hall.

Prügl, R., Schreier, M. (2006). Learning from leading-edge customers at the sims: opening up the innovation process using toolkits. *R&D Management*, **36**, 237–250.

Raasch, C., Herstatt, C., Balka, K. (2009). On the open design of tangible goods. *R&D Management*, **39**, 328–393.

Rau, C., Jonas, J., Schweitzer, F. (2016). Open service prototyping. In: Egger, R., Gula, I., Walcher, D. (eds.) *Open Tourism: Open Innovation, Co-creation Challenges in the Tourism Industry*. Berlin: Springer: pp. 371–381.

Rau, C., Schweitzer, F., Gassmann, O. (2014). Open Foresight Workshops for Opportunity Identification in Open Innovation. In: Noble, C.H., Durmusoglu, S.S., Griffin, A. (eds.) *Open Innovation: New Product Development Essentials from the PDMA*. Hoboken, NJ: John Wiley & Sons Ltd: pp. 27–52.

Reichwald, R., Piller, F., Ihl, C., Seifert, S. (2009). *Interaktive Wertschöpfung: Open Innovation, Individualisierung und neue Formen der Arbeitsteilung*. 2nd Edition. Wiesbaden: Gabler.

Reid, S.E., de Brentani, U. (2004). The fuzzy front end of new product development for discontinuous innovations: a theoretical model. *Journal of Product Innovation Management*, **21**, 170–184.

_____. (2010). Market vision and market visioning competence: impact on early performance for radically new, high-tech products. *Journal of Product Innovation Management*, **27**, 500–518.

_____. (2015). Building a measurement model for market visioning competence and its proposed antecedents: organizational encouragement of divergent thinking, divergent thinking attitudes, and ideational behavior. *Journal of Product Innovation Management*, **32**, 243–262.

Reuer, J.J., Ragozzino, R. (2006). Agency hazards and alliance portfolios. *Strategic Management Journal*, **27**, 27–43.

Reynolds, T.J., Gutman, J. (1988). Laddering theory, method, analysis, and interpretation. *Journal of Advertising Research*, **28**, 11–31.

Ries, E. (2011). *The Lean Startup: How Today's Entrepreneurs Use Continuous Innovation to Create Radically Successful Businesses*. New York, NY: Crown Business.

Rijsdijk, S.A., van den Ende, J. (2011). Control combinations in new product development projects. *Journal of Product Innovation Management*, **28**, 868–880.

Rindfleisch, A., Moorman, C. (2001). The acquisition and utilization of information in new product alliances: a strength-of-ties perspective. *Journal of Marketing*, **65**, 1–18.

Ritter, S. (2012). *Creativity. Understanding and Enhancing Creative Thinking.* Dissertation, Radboud University, Nijmegen, The Netherlands.

Roberts, N., Galluch, P.S., Dinger, M., Grover, V. (2012). Absorptive capacity and information systems research: review, synthesis, and directions for future research. *MIS Quartely*, **36**, 625–648.

Roos, I. (2002). Methods of investigating critical incidents. *Journal of Service Research*, **4**, 193–204.

Rosenfield, D., Folger, R., Adelman, H.F. (1980). When rewards reflect competence: a qualification of the overjustification effect. *Journal of Personality and Social Psychology*, **39**, 368–376.

Rosenkopf, L., Nerkar, A. (2001). Beyond local search: boundary-spanning, exploration, and impact in the optical disc industry. *Strategic Management Journal*, **22**, 287–307.

Rosenthal, S.R., Capper, M. (2006). Ethnographies in the front end: designing for enhanced customer experiences. *Journal of Product Innovation Management*, **23**, 215–237.

Rothwell, R. (1992). Successful industrial innovation: critical success factors for the 1990s. *R&D Management*, **22**, 221–239.

Rubera, G., Kirca, A.H. (2012). Firm innovation and its performance outcomes: a meta-analytic review and theoretical integration. *Journal of Marketing*, **76**, 130–147.

Rushkoff, D. (2007). *Get Back in the Box: How Being Great at What You Do Is Great for Business*. New York, NY: HarperCollins.

Salomo, S., Weise, J., Gemünden, H.G. (2007). NPD planning activities and innovation performance: the mediating role of process management and the moderating effect of product innovativeness. *Journal of Product Innovation Management*, **24**, 285–302.

Schembri, S. (2009). Reframing brand experience: the experiential meaning of Harley–Davidson. *Journal of Business Research*, **62**, 1299–1310.

Schembri, S., Boyle, M.V. (2013). Visual ethnography: achieving rigorous and authentic interpretations. *Journal of Business Research*, **66**, 1251–1254.

Schenk, E., Guittard, C. (2011). Towards a characterization of crowdsourcing practices. *Journal of Innovation Economics*, **7**, 93.

Schirr, G.R. (2012). Flawed tools: the efficacy of group research methods to generate customer ideas. *Journal of Product Innovation Management*, **29**, 473–488.

Schmookler, J. (1966). *Invention and Economic Growth*. Cambridge, MA: Harvard University Press.

Schoormans, J.P.L., Ortt, R.J., de Bont, C.J.P.M. (1995). Enhancing concept test validity by using expert consumers. *Journal of Product Innovation Management*, **12**, 153–162.

Schreier, M., Prügl, R. (2008). Extending lead-user theory: antecedents and consequences of consumers' lead userness. *Journal of Product Innovation Management*, **25**, 331–346.

Schultz, C., Salomo, S., de Brentani, U., Kleinschmidt, E.J. (2013). How formal control influences decision-making clarity and innovation performance. *Journal of Product Innovation Management*, **30**, 430–447.

Schweisfurth, T.G., Herstatt, C. (2015). Embedded (lead) users as catalysts to product diffusion. *Creativity and Innovation Management*, **24**, 151–168.

Schweisfurth, T.G., Raasch, C. (2015). Embedded lead users — the benefits of employing users for corporate innovation. *Research Policy*, **44**, 168–180.

Schweitzer, F., Gassmann, O., Rau, C. (2014). Lessons from ideation: where does user involvement lead us? *Creativity and Innovation Management*, **23**, 155–167.

Schweitzer, F., van den Hende, E. (2017). Drivers and consequences of narrative transportation: understanding the role of stories and domain-specific skills in improving radically new products. *Journal of Product Innovation Management*, **34**, 101–118.

Schweitzer, F., Rau, C., Gassmann, O., van den Hende, E.A. (2015). Technologically reflective individuals as enablers of social innovation. *Journal of Product Innovation Management*, **32**, 847–860.

Schweitzer, F.M. (2013). Integrating customers at the front end of innovation. In: Gassmann, O., Schweitzer, F. (eds.) *Management of the Fuzzy Front End of Innovation*. Cham, Switzerland: Springer: pp. 31–48.

_____. (2014). Market research in the process of new product development. In: Gupta, P., Trusko, B.E. (eds.) *Global Innovation Science Handbook*, 1st Edition. New York, NY: McGraw-Hill: pp. 283–302.

Schweitzer, F.M., Buchinger, W., Gassmann, O., Obrist, M. (2012). Crowdsourcing: leveraging innovation through online idea competitions. *Research-Technology Management*, **55**, 32–38.

Seravalli, A., Simeone, L. (2016). Performing hackathons as a way of positioning boundary organizations. *Journal of Organizational Change Management*, **29**, 326–343.

Sethi, R., Iqbal, Z. (2008). Stage-gate controls, learning failure, and adverse effect on novel new products. *Journal of Marketing*, **72**, 118–134.

Shah, S.K. (2006). Motivation, governance, and the viability of hybrid forms in open source software development. *Management Science*, **52**, 1000–1014.

Shah, S.K., Tripsas, M. (2007). The accidental entrepreneur: the emergent and collective process of user entrepreneurship. *Strategic Entrepreneurship Journal*, **1**, 123–140.

Shaw, N.E., Burgess, T.F., Hwarng, H.B., de Mattos, C. (2001). Revitalizing new process development in the UK fine chemicals industry. *International Journal of Operations & Production Management*, **21**, 1133–1151.

Simon, H.A., Chase, W. (1973). Skill in chess. *American Scientist*, **61**, 394–403.

Slater, S.F., Narver, J.C. (1999). Market-oriented is more than being customer-led. *Strategic Management Journal*, **20**, 1165–1168.

Smith, S., Ward, T., Schumacher, J. (1993). Constraining effects of examples in a creative generation task. *Memory and Cognition*, **21**, 837–845.

Smith, S.M. (2003). The Constraining effects of initial ideas. In: *Group Creativity: Innovation through Collaboration*. Oxford, UK: Oxford University Press: pp. 15–31.

Smith, W.K., Tushman, M.L. (2005). Managing strategic contradictions: a top management model for managing innovation streams. *Organization Science*, **16**, 522–536.

Sørensen, F., Mattsson, J. (2016). Speeding up innovation: building network structures for parallel innovation. *International Journal of Innovation Management*, **20**, 1650024.

Soukhoroukova, A., Spann, M., Skiera, B. (2012). Sourcing, filtering, and evaluating new product ideas: an empirical exploration of the performance of idea markets. *Journal of Product Innovation Management*, **29**, 100–112.

Spann, M., Ernst, H., Skiera, B., Soll, J.H. (2009). Identification of lead users for consumer products via virtual stock markets. *Journal of Product Innovation Management*, **26**, 322–335.

Stanfield, B.R. (2002). *The Workshop Book: From Individual Creativity to Group Action*. ICA series. Toronto, Canada: New Society Publishers.

Stauss, B., Weinlich, B. (1997). Process-oriented measurement of service quality: applying the sequential incident technique. *European Journal of Marketing*, **31**, 33–55.

Stein, M.I. (1953). Creativity and culture. *The Journal of Psychology: Interdisciplinary and Applied*, **36**, 311–322.

Sullivan, B. (2010). Competition and beyond: problems and attention allocation in the organizational rulemaking process. *Organization Science*, **21**, 432–450.

Sutherland, J. (2015). *Scrum: The Art of Doing Twice the Work in Half the Time*. London, UK: Random House Business.

Taggar, S. (2002). Individual creativity and group ability to utilize individual creative resources: a multilevel model. *The Academy of Management Jounrnal*, **45**, 315–330.

Talkea, K., Salomob, S., Rost, K. (2010). How top management team diversity affects innovativeness and performance via the strategic choice to focus on innovation fields. *Research Policy*, **39**, 907–918.

Tan, F.B., Hunter, M.G. (2002). The repertory grid technique: a method for the study of cognition in information systems. *MIS Quarterly*, **26**, 39–57.

Tax, S.S., Stuart, F.I. (1997). Designing and implementing new services: the challenges of integrating service systems. *Journal of Retailing*, **73**, 105–134.

Techcrunch. (2013). Available at https://techcrunch.com/2013/11/21/two-harvard-university-alum-win-salesforce-1m-hackathon-prize-at-dreamforce-for-mobile-service-to-create-reports/ (accessed 30 August 2017).

Terwiesch, C., Xu, Y. (2008). Innovation contests, open innovation, and multi-agent problem solving. *Management Science*, **54**, 1529–1543.

Thomas, D.M., Bostrom, R.P. (2010). Vital signs for virtual teams: an empirically developed trigger model for technology adaptation interventions. *MIS Quarterly*, **34**, 115–142.

Thomke, S. (2014). Accelerating learning by experimentation. In: Gassmann, O., Schweitzer, F. (eds.) *Management of the Fuzzy Front End of Innovation*. Cham, Switzerland: Springer: pp. 125–140.

Thomke, S.H., Nimgade, A. (1996). Note on lead user research. In Burgelman, R.A., Christensen, C.M., Wheelwright, S.C. (eds.) *Strategic Management of Technology and Innovation*. Burr Ridge, IL: Irwin/McGraw-Hill.

Tidd, J. (2014). *Open Innovation Research, Management and Practice*. London, UK: Imperial College Press.

Tidd, J., Bessant, J. (in press). *Managing Innovation: Integrating Technological, Market and Organizational Change*. 6th Edition. Chichester: John Wiley & Sons.

Tidd, J., Hull, F.M. (2003). *Service Innovation: Organizational Responses to Technological Opportunities and Market Imperatives*. London, UK: Imperial College Press.

_____. (2006). Managing service innovation: the need for selectivity rather than 'best-practice'. *New Technology, Work and Employment*, **21**, 139–161.

Tideman, S. (2015). *Business as an Instrument for Societal Change: In Conversation with the Dalai Lama*. Shipley, UK: Greenleaf Publishing.

Toubia, O. (2006). Idea generation, creativity, and incentives. *Marketing Science*, **25**, 411–425.

Tuckel, P., Wood, M. (2001). Respondent cooperation in focus groups: a field study using moderator ratings. *International Journal of Market Research*, **43**, 391–407.

Ulwick, A. (2005). *What Customers Want: Using Outcome-Driven Innovation to Create Breakthrough Products and Services*. New York, NY: McGraw-Hill.

Unilever. (2015). Unilever Sustainable Living Plan.

Urban, G.L., Hauser, J.R., Quails, W.J., Weinberg, B.D., Bohlmann, J.D., Chicos, R.A. (1992). Premarket forecasting of really-new products. *Journal of Marketing Research*, **34**, 143–153.

Urban, G.L., Hauser, J.R., Qualls, W.J., Weinberg, B.D., Bohlmann, J.D., Chicos, R.A. (1997). Information acceleration: validation and lessons from the field. *Journal of Marketing Research*, **34**, 143–153.

Urban, G.L., von Hippel, E. (1988). Lead user analyses for the development of new industrial products. *Management Science*, **34**, 569–582.

Urban, G.L., Weinberg, B.D., Hauser, J.R. (1996). Premarket forecasting of really-new products. *Journal of Marketing*, **60**, 47–60.

Utterback, J.M. (1971). The process of technological innovation within the firm. *Academy of Management Journal*, **14**, 75–88.

Utterback, J.M., Abernathy, W.J. (1975). A dynamic model of process and product innovation. *Omega*, **3**, 639–656.

Vahs, D., Brehm, A. (2013). *Innovatonsmanagement: Von der Idee zur erfolgreichen Vermarktung*. 4th Edition. Stuttgart, Germany: Schäffer-Poeschel.

Verganti, R. (1999). Planned flexibility: linking anticipation and reaction in product development projects. *Journal of Product Innovation Management*, **16**, 363–376.

_____. (2016). *Overcrowded: Designing Meaningful Products in a World Awash with Ideas*. Cambridge, MA: MIT Press.

Veryzer, R.W. (1998). Key factors affecting customer evaluation of discontinuous new products. *Journal of Product Innovation Management*, **15**, 136–150.

Walker, M. (2002). Customer-driven breakthroughs using QFD and policy deployment. *Management Decision*, **40**, 248–256.

Weiss *et al.* (2008). Listening to strangers: whose responses are valuable, how valuable are they and why? *Journal of Martketing Research*, **45**, 425–436.

Wellner, K., Herstatt, C. (2014). Determinants of user innovator behaviour in the silver market. *International Journal of Innovation Management*, **18**, 1–24.

Welsh, G.S. (1975). *Creativity and Intelligence*. Chapel Hill, NC: Institute for Research in Social Science, University of North Carolina Press.

Whyte, J., Ewenstein, B., Hales, M., Tidd, J. (2008). How to visualize knowledge in project-based work. *Long Range Planning*, **41**, 74–92.

Wikipedia. (2017). Kinect. Available at https://en.wikipedia.org/wiki/Kinect#Open_source_drives (accessed 30 August 2017).

Williamson, O.E. (1985). *The Economic Institutions of Capitalism*. New York, NY: Free Press.

Wirdemann, R., Mainusch, J. (2017). *Scrum mit User Stories*. 3rd Edition. Krugzell, Germany: Carl Hanser.

Xie, M., Goh, T.N., Tan, K.C. (2003). *Advanced QFD Applications*. Milwaukee, WI: ASQ Quality Press.

Zaichkowsky, J.L. (2013). The personal involvement inventory: reduction, revision, and application to advertising. *Journal of Advertising*, **23**, 59–70 (Research Notes).

Zaltman, G. (2003). *How Customers Think: Essential Insights in the Mind of the Market*. Boston, MA: Harvard Business School Press.

Zaltman, G., Zaltman, L. (2008). *Marketing Metaphoria: What Deep Metaphors Reveal about the Minds of Consumers*. Boston, MA: Harvard Business School Press.

Zeithaml, V., Bitner, M.J., Gremler, D. (2006). *Services Marketing: Integrating Customer Focus Across the Firm*. 4th Edition. Boston, MA: McGraw-Hill.

Zhao, M., Hoeffler, S., Dahl, D.W. (2009). The role of imagination-focused visualization on new product evaluation. *Journal of Marketing Research*, **46**, 46–55.

Zhao, M., Hoeffler, S., Dahl, D.W. (2012). Imagination difficulty and new product evaluation. *Journal of Product Innovation Management*, **29**, 76–90.

Zhao, M., Zauberman, G. (2011). Mental simulation and product evaluation: the affective and cognitive dimensions of process versus outcome simulation. *Journal of Marketing Research*, **48**, 827–839.

Index

Series on Technology Management

(Continuation of series card page)

Vol. 24 *The Knowledge Enterprise: Innovation Lessons from Industry Leaders (Second Edition)*
by E. Huizenga (University of Amsterdam, The Netherlands)

Vol. 23 *Open Innovation Research, Management and Practice*
edited by J. Tidd (University of Sussex, UK)

Vol. 22 *Discontinuous Innovation: Learning to Manage the Unexpected*
by P. Augsdörfer (Technische Hochschule Ingolstadt, Germany),
J. Bessant (University of Exeter, UK),
K. Möslein (Universität Erlangen-Nürnberg, Germany),
B. von Stamm (Innovation Leadership Forum, UK) &
F. Piller (RWTH Aachen University, Germany)

Vol. 21 *Workbook for Opening Innovation:Bridging Networked Business, Intellectual Property and Contracting*
by Jaakko Paasi (VTT Technical Research Centre of Finland, Finland),
Katri Valkokari (VTT Technical Research Centre of Finland, Finland),
Henri Hytönen (VTT Technical Research Centre of Finland, Finland),
Laura Huhtilainen (University of Eastern Finland, Finland) &
Soili Nystén-Haarala (University of Eastern Finland, Finland)

Vol. 20 *Bazaar of Opportunities for New Business Development: Bridging Networked Innovation, Intellectual Property and Business*
by Jaakko Paasi (VTT Technical Research Centre of Finland, Finland),
Katri Valkokari (VTT Technical Research Centre of Finland, Finland),
Tuija Rantala (VTT Technical Research Centre of Finland, Finland),
Soili Nystén-Haarala (University of Eastern Finland, Finland),
Nari Lee (University of Eastern Finland, Finland) &
Laura Huhtilainen (University of Eastern Finland, Finland)

Vol. 19 *From Knowledge Management to Strategic Competence:Assessing Technological, Market and Organisational Innovation (Third Edition)*
edited by Joe Tidd (University of Sussex, UK)

Vol. 18 *Perspectives on Supplier Innovation: Theories, Concepts and Empirical Insights on Open Innovation and the Integration of Suppliers*
edited by Alexander Brem (University of Erlangen-Nuremberg, Germany) &
Joe Tidd (University of Sussex, UK)

Vol. 17 *Managing Process Innovation: From Idea Generation to Implementation*
by Thomas Lager (Grenoble Ecole de Management, France)

Vol. 16 *Perspectives on User Innovation*
edited by Stephen Flowers (University of Brighton, UK) &
Flis Henwood (University of Brighton, UK)

Vol. 15 *Gaining Momentum: Managing the Diffusion of Innovations*
edited by Joe Tidd (SPRU, University of Sussex, UK)

Vol. 14 *Innovation and Strategy of Online Games*
by Jong H. Wi (Chung-Ang University, South Korea)

Vol. 13 *Building Innovation Capability in Organizations: An International Cross-Case Perspective*
by Milé Terziovski (University of Melbourne, Australia)

Vol. 12 *Project-Based Organization in the Knowledge-Based Society*
by Mitsuru Kodama (Nihon University, Japan)

Vol. 11 *Involving Customers in New Service Development*
edited by Bo Edvardsson (Karlstad University, Sweden),
Anders Gustafsson (Karlstad University, Sweden),
Per Kristensson (Karlstad University, Sweden),
Peter Magnusson (Karlstad University, Sweden) &
Jonas Matthing (Karlstad University, Sweden)

Vol. 10 *Open Source: A Multidisciplinary Approach*
by Moreno Muffatto (University of Padua, Italy)

Vol. 9 *Service Innovation: Organizational Responses to Technological Opportunities & Market Imperatives*
edited by Joe Tidd (University of Sussex, UK) &
Frank M. Hull (Fordham University, USA)

Vol. 8 *Digital Innovation: Innovation Processes in Virtual Clusters and Digital Regions*
edited by Giuseppina Passiante (University of Lecce, Italy),
Valerio Elia (University of Lecce, Italy) &
Tommaso Massari (University of Lecce, Italy)

Vol. 7 *Innovation Management in the Knowledge Economy*
edited by Ben Dankbaar (University of Nijmegen, The Netherlands)

Vol. 6 *Social Interaction and Organisational Change: Aston Perspectives on Innovation Networks*
edited by Oswald Jones (Aston University, UK),
Steve Conway (Aston University, UK) & Fred Steward (Aston University, UK)

Vol. 5 *R&D Strategy and Organisation: Managing Technical Change in Dynamic Contexts*
by Vittorio Chiesa (Università degli Studi di Milano-Bicocca, Milan, Italy)

Vol. 4 *Japanese Cost Management*
by Yasuhiro Monden (University of Tsukuba, Japan)

Vol. 3 *From Knowledge Management to Strategic Competence: Measuring Technological, Market and Organizational Innovation*
edited by Joe Tidd (SPRU, University of Sussex, UK)

Vol. 2 *The Knowledge Enterprise: Implementation of Intelligent Business Strategies*
by J. Friso den Hertog (MERIT, Maastricht University & Altuïtion bv, 's Hertogenbosch, The Netherlands) &
Edward Huizenga (Altuïtion bv, 's Hertogenbosch, The Netherlands)

Vol. 1 *Engines of Prosperity: Templates for the Information Age*
by Gerardo R. Ungson (University of Oregon, USA) &
John D. Trudel (The Trudel Group, USA)

www.ingramcontent.com/pod-product-compliance
Ingram Content Group UK Ltd.
Pitfield, Milton Keynes, MK11 3LW, UK
UKHW020348150926
8624IPUK00004B/65

9 781786 345363